MY PERSONAL HISTORY OF THE BRITISH INTERPLANETARY SOCIETY

LIVERPOOL 1933 TO 1937

LESLIE J JOHNSON   PAM REID   GURBIR SINGH

ASTROTALKUK PUBLICATIONS

Copyright © 2021 Leslie J Johnson

All rights reserved. No part of this publication may be reproduced, distributed or transmitted in any form or by any means, including photocopying, recording, or other electronic or mechanical methods, without the prior written permission of the publisher and estate of Leslie J Johnson.

Exceptions include brief quotations embodied in critical reviews and specific other noncommercial uses permitted by copyright law. For permission requests, contact the publisher at the email address below.

Gurbir Singh/Astrotalkuk Publications

www.astrotalkuk.org

info@astrotalkuk.org

Front and back cover images adapted from the Leslie J Johnson collection.

ISBN: 978-1-913617-07-3

My Personal History of

The British Interplanetary Society

1933 - 1937 Liverpool

Leslie J Johnson (18 May 1914 - 3 July 1982)

Manuscript completed by Leslie Johnson during the late 1970s and early 1980s

---

With

Foreword by Pam Reid

Introduction and Epilogue by Gurbir Singh

FOREWORD BY PAM REID

This book comprises my father's recollections as a founder member of the British Interplanetary Society. It covers details of his younger life leading on to the first meeting of the BIS on 13 October 1933. He wrote it in the late 1970s, early 1980s with the 50th anniversary of the first meeting in mind. Sadly, he passed away on 3 July 1982 so didn't live to see the fruition of his work. Very few of his contemporaries from that time other than Arthur Clarke and, as far as I know, Colin Askham were alive in 1983.

A sixteen-year-old Arthur Clarke wrote to dad in 1934 asking if he could have particulars of the Society with a view to joining it. By 1968 he was a famous science fiction enthusiast and writer. His film "2001: A Space Odyssey" was on at the local cinema. I attended a showing with my mum and sister. Dad preferred to wait until the tickets Arthur had promised to him arrived from Metro Goldwyn Mayer. Arthur visited our family home years before I was born. We have a signed photograph taken in June 1937 and dedicated 'To Les'.

I can remember visiting Colin Askham's house in Formby sometime around the late 1960s. Colin seemed old to a girl in her twenties! A bird had died in his garden and he was quite upset about this. A few years after that the four of us – mum, dad and my sister Val and I stayed the weekend in Buxton where there was a science fiction convention. Norman Weedall joined in the fancy dress. He was a founder member, first Hon Librarian and remained friends with dad all of his life.

My earliest memories of my father's interest in science fact and fiction were having a lot of science fiction books and magazines in the house. As a child, I used to 'help' him to parcel them up to be posted abroad. As I grew older my mum used to tell me how everyone thought that, in the 1930s, she was engaged to a crank as most people thought that talking about people being able to fly to the moon was crazy! This came as a surprise to me because having grown up with the idea all my life, it did not seem at all strange.

I spent the night of 20 July 1969 in my newly married sister's flat in Hoylake, on the Wirral. In common with most people, I was awake most of the night waiting to see the lunar landing on the television. My dad, on the other hand, having spent the afternoon at Colin Askham's house, watched it on a small television in his bedroom that evening, falling asleep at about 12.30 am. He happened to wake up five minutes before Neil Armstrong emerged from the lunar module, almost as if some instinct had guided him. Those were the days before you could record programmes so if he had missed it the only chance to see it again would be on the news.

I married and left home in 1979 and gave birth to my first child exactly a year later. As I only moved two roads away I often used to take my son Martin round to my parents' house in the pram. I can remember many times, my dad busily typing away in the front room.

As dad mentioned in his transcript he made many lifelong friends through the BIS with people who had the same interest.

Many thanks to Gurbir Singh for all his interest, help and encouragement that made this book possible.

Pam Reid - Liverpool October 2020

INTRODUCTION BY GURBIR SINGH

By chance, I came across this manuscript from Leslie Johnson in 2019 whilst visiting Pam Reid (Leslie Johnson's daughter) in Liverpool.[1] I was then researching the work of a rocket mail experimenter in India called Stephen Smith.[2] As the BIS's first honorary secretary in 1933, Johnson was the first point of contact for enquiries that came in from around the world. His accumulated correspondence became a treasure trove that captured a unique account of the early stages of a pivotal technology emerging at a critical time in world history.

Leslie Johnson wrote the manuscript between 1974 and 1979 after retiring as an administrator from the Liverpool Education Offices. He records the BIS story during the Liverpool years, specifically from its founding in 1933 to 1939. This first-hand account of a global spaceflight movement by an unassuming young man from Liverpool should have been published four decades ago. Given how deeply our everyday lives depend on space technology today, this account of how this spaceflight movement emerged in 1930s Liverpool is even more pertinent now. A draft was completed in 2020, but the Covid 19 pandemic delayed the final version with the epilogue to April 2022.

The manuscript describes in detail the BIS story from the day it was founded on 13 October 1933 to a handover to the London Branch on 7th February 1937. Following a recommendation from Arthur C. Clarke, who described the manuscript as "fascinating" and "a valuable historical document", in the autumn of 1980, Johnson sent a copy to the National Air and Space Museum in Washington DC.[3] Johnson was seeking to publish it to mark the 50th anniversary of the BIS in 1983 but sadly he passed away on 3 July 1982. It has remained unpublished since.

In this first-hand account, Leslie Johnson captures the events and the people who founded the BIS. He describes the hurdles in the BIS's early history and looking back from this vantage point, I can see more. The difficulties they faced included the Explosives Act of 1875; the increasing risk-averse stance of the British government following the disastrous R101 failure in 1930, Gerhardt Zucker's spectacular rocket launch failures in Britain, and Phil Cleator's communication style perceived by many as "dictatorial". Despite this headwind, remarkable progress was made by the BIS during the prewar days in Liverpool.

I had planned to conclude the BIS story in a single chapter that summarised the post-war activities. I intended to follow up the story from where Johnson had left it. I wanted to understand how the war had impacted the technology of rockets and shaped the lives of some of the founding members. World War Two was a turning point. It dramatically accelerated the technology of rockets and the mindset of those who had championed it. The story turned out to be a richer and more complex one that cannot be told in a single chapter. So a separate book, *"From Imagination to Reality: Founding of the British Interplanetary Society"* will be published in 2022. It is based on my research and experience of being a member of the BIS for over a decade. The most comprehensive and authoritative source of the BIS history remains *"Interplanetary. A History of the British Interplanetary Society"* published in 2008, edited by Robert Parkinson.[4] Most of its contributions are directly from BIS members who played a significant role within the BIS when it was most influential.

Apart from the minor contributions from me (Introduction and Epilogue) and Leslie Johnson's daughter (Pam Reid), this work is that of Leslie Johnson that he compiled between 1974 and 1979. In the final chapter, the epilogue, I attempted to understand from a 21st-century perspective, the BIS's initial objectives, the hurdle of convincing a population that was just coming to terms with the idea of air travel that the phenomenon of space and space travel was a reality. A reality that was accessible using prevailing technology. Until then, every mode of transport needed something to "push against". In the vacuum of space, there was nothing to push against. The destructive V2 rockets developed and used during World War Two, demonstrated that rockets could work in a vacuum. This was a frequent conceptual challenge that troubled the sceptical public during the 1930s. Leslie Johnson used "*Nothing to Push Against*" as the working title for this manuscript.

I want to thank several individuals who have helped with my research including Pam Reid, Robert Parkinson, Grif Ingram, John Harlow, John Davies, Brian Harvey and several librarians at the British Library. Despite the cycle of proofreading and checks, some errors will have crept through. The responsibility for them is entirely mine. If you spot any - please drop me an email.

---

[1] Not to be confused with Les Johnson - the NASA technologist. https://www.lesjohnsonauthor.com/

[2] Singh, G. 2019, India's Forgotten Rocket Pioneer. From Pigeon Mail to Rocket Mail"

[3] Following his death in 2008, Arthur C Clarke's archive ended up at the National Air and Space Museum. His brother Fred Clark maintains another smaller archive in Taunton, Somerset.

[4] The BIS will be publishing an updated version in 2023 - the 90th anniversary of the BIS.

A popular opinion in 1930 to 1940

"A rocket cannot work in the vacuum of space - because there would be no air for it to push against"

DEDICATION

To those who did so much towards the foundation of the British Interplanetary Society

Liverpool

Philip E Cleator (First President of the Society)

Colin H Askham (First Vice President)

James A Free, Eric Frank Russell and Percival Norman Weedall

London

Edward John Carnell, Arthur C Clarke, J Happian Edwards, Walter H Gillings, Morris K Hanson and William F Temple

ABOUT LESLIE JOHNSON

First Hon Secretary of the British Interplanetary Society, 13 October 1933

Founder Fellow of the British Interplanetary Society, 1933 to 1945

Editor the *Bulletin* of the British Interplanetary Society, 1933 to 1934

Joint editor with Eric Frank Russell of the *New Columbus*, a publication of the British Interplanetary Society, January 1935

Editor the *Journal* of the British Interplanetary Society, 1936 to 1937

Hon Secretary-Treasurer of the British Interplanetary Society, 1936

Hon General Secretary-Treasurer of the British Interplanetary Society, 1937

Vice President of the British Interplanetary Society, 1937 to 1945

Co-author with Eric Frank Russell of:

- *Seeker of Tomorrow* (*Astounding Stories* July 1937 - recently anthologised in Britain, USA and Spain)
- *Eternal Rediffusion* (*Fantasy Booklet* No.3 and *Weird Tales* Fall 1973 - recently anthologised in Britain and Italy)
- Author of: *Satellites of Death* (*Tales of Wonder* No.3)
- *Transatlantic Rocket Mails* (*Meccano Magazine* September 1937)
- *Ahead of Reality* (*New Worlds* No. 2)

Contributor to pre-war Science Fiction fan magazines

Council member of the Science Fiction Association, 1937 to 1939

Partner with Edward John Carnell in Science Fiction Service, 1937 to 1945

Editor of the *Science Fantasy Review 1939*.

Editor of *Outlands* Winter 1946

PREFACE

I was 19 years of age when the British Interplanetary Society was founded and held its first meeting on 13 October 1933; having been the only volunteer for the post, I was duly accepted as the Society's first Hon Secretary. Considering the number of persons present at that first meeting failed to reach double figures, and that I owned my own typewriter - and knew how to use it, the "honour" of having been appointed as Hon Secretary of the new organisation was probably not as profound as it may have seemed.

However, it may be of some interest, if I also volunteer some information as to my personal background that one way and another led me to be the Hon Secretary of the Society.

In 1926, at the age of 12 years, I gained a Free Place Scholarship which entitled me to the benefit of being educated by the Jesuits at St Francis Xavier's College in Liverpool. The standards set at the College were incredibly high; only to have gained a School Certificate was regarded as an arrant failure, so that when I left the College at the age of 16 years in mid - 1930, apart from the inevitable School Certificate, I was awarded the Matriculation Certificate of the Northern Universities.

I also gained distinction in Latin and mathematics, good in chemistry, credits in French, history and physics and a pass (only) in English composition (in which I wrote an essay on "A City of the Future"). Subsequently, the fact that I could not achieve better than a pass in English precluded me from being accepted for a course in the London External BSc at the Byrom Street Technical College, Liverpool; without having obtained a credit at least in this subject my Northern Universities Matriculation Certificate (no matter what I might have achieved in other subjects) was not considered to be the equivalent of a London Matriculation.

I was advised to take English again as a separate subject in an endeavour to obtain the required credit rating. But this I refused to do.

When I left the College, there were 3,000,000 unemployed in Britain, and there was no such thing as social security payments. Anyone who was a destitute had to go cap in hand to "The Parish Council" and plead for a handout, we would rather have died than to have done this. In May 1923, my father had retired as a Band Sergeant after 23 years with the First King's (Liverpool) Regiment. Although a very competent musician and an Army school teacher, the only job he could get was cleaning tram cars at the Edge Lane depot. Fortunately, he also had a pension from the Army. We lived at 46 Mill Lane, Old Swan, Liverpool 13, which was a sweets and tobacconist shop, which my mother herself had bought with money she had inherited from her own mother.

I was without employment from July 1930 until January 1931, when I obtained a post as a junior clerk at the School Medical Department at the Liverpool Educational Committee. The day on which I should have started in the School Medical Department, Monday 12 January 1931, I went to my father's funeral instead. He had caught pneumonia while playing the euphonium with the Tramways Band at a party given for children at St George's Hall, Liverpool, which he had complained had been cold and draughty. He was only 50 years of age. My mother survived him by over 45 years, dying suddenly during the night of Thursday, 19 August 1976, only two months from her 90th birthday.

I therefore took up my duties as a clerk on Tuesday, 13 January 1931 at an annual salary of forty-five pounds. I found that one of my colleagues was J Free Jr. who I had already met during interviews for the posts held during the previous October. Jimmy, who was of a similar age to myself, was to become one of the Founder Fellows of the British Interplanetary Society.

In order to further my career in the Liverpool Education offices (where all-male staff were expected to be able to do all their own shorthand and typewriting), my mother paid five pounds for a Smith Premier Typewriter - a model which was obsolete even then. It had a double bank keyboard, no shift keys and with separate keys for all upper and lower case letters as well as for figures and miscellaneous symbols. I was able to master its intricacies, however, and thus became qualified and became the first Hon Secretary of the BIS.

Jimmy had started his job three months previously and I found I had in fact been sent for to take up my post in a hurry because one of the clerks of the School Medical Department had been killed in a car accident on Christmas Eve, 1930. Jimmy and myself alternated weekly between operating the post desk and writing out dental notices. These notices were sent to parents, advising them that their children needed dental treatment, which was available from the school dentists at a cost of 6d. (2½p). Should the parents be unable to afford this amount treatment might be provided free of charge.

I had the experience of nearly three years in this work, and was a competent typist when the time came for the formation of the British Interplanetary Society. Although I studied both Pitman's and Gregg's shorthand, I could never do any good in either.

From an early age, I had shown an interest in the more imaginative types of literature, including stories concerning Beowulf, then fiction by A. Conan Doyle, Jules Verne and HG Wells. There had been stories of mechanical armies in *Comic Cuts*, while I used to look forward eagerly to reading the *Boys Magazine* every Saturday morning, a publication which regularly featured stories of a kind, we now know as science fiction. During the summer of 1930, when I had left SFX

College, was unemployed, and had more time to look around the book shops of Liverpool, I came across the issue of the American science fiction magazines. I also became an ardent collector of second-hand books on astronomy.

Looking back at my work for the British Interplanetary Society. I do so with the greatest satisfaction. I am staggered when I consider the vast amount I must have got through. First, as Hon Secretary later as Hon Treasurer then as Hon General Secretary-Treasurer -not to mention having edited the *Bulletin,* the *New Columbus* and the *Journal* of the Society.

Over 45 years have elapsed since the foundation of the BIS in October 1933, and the names of the original members of the Society (both in Liverpool and in London) are fading with the passage of time. It was with this thought in mind that I have written this account of the Society in its earliest years.

The crux of the situation was that from the beginning, we had aspired to be the "British" Interplanetary Society, a national, if not an international organisation. We were not to be the Liverpool Interplanetary Society. Subsequent to the formation of the BIS, other societies with similar objectives sprung up in various parts of the country, but none of them at that time made any pretence towards being other than a local organisation. Had we been content to have been a parochial institution it is very doubtful whether we would have experienced the difficulties that we did. It is also very doubtful whether the Society would have survived to the present day.

Amongst the original members on Merseyside, every tribute must be paid to P Cleator, Founder and the first President of the Society, as well as to Colin Askham, the first Vice President. Mention must be made of the valuable contributions made by James Free Jr as Assistant Hon Secretary, and as Hon Treasurer, as well as my late friend, Normal Weedall (1914-1978) as Hon Librarian. Particular mention must be made of the advice, efforts and contributions of science fiction writer Eric Russell (1905- 1978) during the formative years of the Society and the very early days of his successful writing career.

Amongst the early London members, during the period of 1936 to 1939, the efforts and influence of J Happian Edwards, founder of the London Branch of the Society must be recorded, together with the secretarial assistance of Miss Elizabeth Huggett (later to become Mrs J.H. Edwards). Not the least of the London members, later to become President of the Society, was Professor AM Low D Sc. Following the transfer of the Society's headquarters from Liverpool to London, in February 1937 few could have worked harder to maintain the development of the Society than Edward Carnell (1912-1972) and Arthur Clarke, now, a well-known resident of Sri Lanka. Other names amongst the Londoners who gave invaluable assistance to the pre-war Society were Walter Gillings, William Temple, Ralph Smith, H Ross, Arthur Janser, H Bramhill and F Day.

Apart from having assisted in the foundations of a scientific Society that has survived for nearly 50 years and that has seen its incredible major objective (the conquest of space) achieved during that time, the existence of the Society more than justified itself in a much more mundane manner - the establishment - over the course of the years of many mutual and lifelong friendships. This is in spite of difficulties that were encountered and the differences of opinion that existed.

Unfortunately, the number of early members who still survive is diminishing year by year. Edward Carnell passed away in 1972 and 1978 saw the "Eternal Rediffusion" of Eric Russell and Norman Weedall, and in 1979 the passing of Walter Gillings.

The British Interplanetary Society is even now preparing to celebrate its 50th anniversary in 1983. One wonders how many of the earliest members from both Liverpool and London will still be available to join the celebration on 13 October 1983...

(*Eternal Rediffusion* - a short story collaboration by Eric Russell and Leslie Johnson concerning life after death and then the meaning and the purpose of existence, written in 1935 and first published nearly 40 years later).

Summary of events

- At 19 years of age the first Hon Secretary of the BIS
- In the mid 1930s - leaving St Francis Xavier's College Liverpool, with the "Metric" a pass only in English composition disqualifies an attempt at the BSc
- Family background
- To father's funeral, instead of starting employment
- Starting life as a junior clerk together with Mr James Free Junior
- Learning to type on a typewriter with a double-bank keyboard
- Work as a junior clerk in the School Medical Department
- An interest at an early age in science fiction and astronomy
- A satisfactory association with the BIS
- In memory of the founders of the BIS
- The BIS - a national institution
- Tribute to the early Liverpool members
- Establishment of lifelong friendship
- The 50th anniversary of the BIS on 13 October 1983

THE YEAR 1933

> *The British Interplanetary Society was "Founded for the stimulation of public interest in the possibility of interplanetary travel and to promote research in all problems pertaining to the conquest of space with the aid of the rocket motor or by any other means."*

Constitution of the Society, 1936.

> *The first meeting of the BIS was held at the office of HC Binns Esq Room 15 on the second floor of No. 81, Dale Street, Liverpool, starting at 7 pm on Friday 13 October 1933.*

I am writing the history of the British Interplanetary Society, covering the years 1933 to 1937 when the headquarters of the Society was based upon Merseyside. I am very fortunate inasmuch as I have preserved items of personal correspondence, documents, press cuttings, and various publications relating to science fiction, and interplanetary travel, going back as far as the year 1930. Without the information that's available to me this narrative could never had been written.

Prior to the foundation of the BIS, so called interplanetary societies had previously existed in Russia (in the first quarter of the century) in Germany (since 1927) and in the United States of America.

I have a cutting from the American science fiction magazine *Wonder Stories Quarterly* dated Summer 1930 announcing the formation of the American Interplanetary Society. This letter was from CP Mason the first Secretary of the Society, who explained that amongst their members was Dr. Robert H Goddard (even in those days an experimenter with liquid fuel rockets) of Clark University, Dr Clyde Fisher of the American Museum of Natural History and Captain Sir Hubert Wilkins the noted explorer, as well as Hugo Gernsback. Hugo Gernsback, ("the father of science fiction") was the editor of *Wonder Stories Quarterly*, as well as being the editor of *Science Wonder Stories, Air Wonder Stories and Scientific Detective Monthly*. He had earlier been editor and publisher of the American magazine, *Amazing Stories*, the first all-science-fiction magazine to have seen publication in the English language, the initial issue having been dated April 1926.

The officers of the American Interplanetary Society had included David Lasser (author of a book entitled *The Conquest of Space*, one of the earliest books on the possibility of interplanetary travel to have appeared in the English language), as President of the Society. As Vice President was G Edward Pendray (author of *The Coming Age of Rocket Power* and *Men, Mirrors and Stars*). Pendray had also had magazine Science-Fiction published under the pseudonym "Gawain Edwards". Laurence Manning was Treasurer of the Society and Fletcher Pratt was Librarian. Laurence Manning and Fletcher Pratt were renowned in the thirties for their collaboration in the writing of science fiction stories in Science Wonder Stories (later to be re-entitled simply *Wonder Stories.*

Science Wonder Stories had a slogan in those days (possibly coined by Gernsback), that read "prophetic fiction is the mother of all scientific fact". David Lasser had been the managing editor of several of Gernsback's science fiction magazines while CP Mason had been his associate editor. A link between science fiction magazines and interplanetary travel had been established in the earliest days of both movements, not only by the publication of stories concerning interplanetary travel, but also by the publication of serious articles on the possibilities of travel in space. Typical of such articles was the publication of The Problem of Space Flying by Captain Hermann Noordung, a Berlin scientist. The narrative appeared in the July, August and September 1929 issues of *Science Wonder Stories*.

On reflection, it seems hardly surprising that most of the advocates of interplanetary travel at the time were science fiction fans and this certainly applies, as far as Britain and America were concerned. In Germany, it is possible that a higher ratio existed of actual scientists and technicians, rather than of dreamers. Science fiction fans inevitably believed in the possibility of interplanetary travel but strangely enough, those scientists who believed in the possibility of interplanetary travel did not always have an interest in science fiction.

The prime advocate in Britain of interplanetary travel was a young scientist - Philip Ellaby Cleator, from an address on the South Bank of the River Mersey in Wallasey, then in the county of Cheshire but now, in the Wirral District of Merseyside County Council. Cleator aspired more to popular scientific journalism than to science fiction. He did, however, contribute a story called *Martian Madness* to the March 1934 issue of *Wonder Stories*, and a serial, *Mutiny on the Moon* published in the Daily Express as shall be explained shortly.

More importantly, in the January 1933 issue of Chambers' Journal, he had published an article entitled "The Possibilities of Interplanetary Travel". According to Cleator's book *Rockets Through Space* (to be published in 1936) it would appear that at the time of the publication of his article in Chambers' Journal he did in fact have in mind the idea of the formation in Britain of an interplanetary society. However, in the article, for whatever reason, he did not make any appeal for members for such a society but explained and examined the difficulties surrounding what may be called the Conquest of Space.

There the matter rested until August 1933. Cleator then saw an opening, when an account appeared in the Liverpool Echo concerning the prediction of one WA Conrad, of the United States Naval Academy regarding the possibility of a journey by rocket to the Moon.

Cleator wrote a letter to the editor of the Liverpool Echo enclosing a copy of his article in Chambers' Journal; this bought a visit from a representative, so that following his interview with Cleator it was arranged the Liverpool Echo would print an appeal for members of the proposed Society. Accordingly, the following letter appeared in the Liverpool Echo on 8 September 1933:

British "Rocketeers"

Reaching the Moon - and Elsewhere

To the editor of the Echo:

There appeared in the Echo recently a brief account of a statement made by Mr WA Conrad, of the United States Naval Academy, on the possibility of reaching the Moon by rocket.

It is significant to note that Mr Conrad is an American.

Few people in this country realise the tremendous strides that have been made in the United States toward the solution of the problem of space travel. There, the immense possibilities that lie dormant in the rocket are widely appreciated. More important still, they are being exploited with an energetic enthusiasm that can only be described as characteristic.

Three years ago, a group of interested experimentalists founded the American Interplanetary Society, under the leadership of that famous pioneer, Dr. Goddard. Its numerous members are popularly known as rocketeers, and their invaluable researches have raised widespread interest. Hard-headed businessmen have given liberal financial aid - the late Daniel Guggenheim to the extent of £20,000 a fact which speaks for itself.

Then there is Germany, whose scientists have an enviable record the world over. She, too, has an interplanetary Society. France and Russia are also deeply interested in the question. England is years behind. So far as I know, the problems of interplanetary travel have received little or no real attention here as yet. Apart from a few isolated experimenters, such as myself, who probably lack the means to carry out research as they would wish, there appears to be nothing being done.

Before me as I write is a letter, I have recently received from Mr CP Mason, secretary of the American Interplanetary Society. He says "... feeling that there is a field for a British (Interplanetary) Society, *which we hope to see organised at the earliest moment*". The italics are mine. The immediate formation of the British Interplanetary Society is imperative if we are to keep in pace with other progressive nations.

Those of us who are interested in the fascinating problems which surround the conquest of space must band together and pool our knowledge in order to ensure the cooperation that leads to success.

Enthusiasts may be subject to derision. This atavistic attitude has always existed, and probably always will exist. Only yesterday the aeroplane and wireless telegraphy were objects of scorn. Now it is this, and that, and interplanetary travel. Tomorrow, for a certainty, there will be other revolutionary suggestions which will offend the unimaginative mind. Perhaps those interested will communicate with me? - P. E. Cleator, 34, Oarside Drive, Wallasey, Cheshire.

My colleagues in the School Medical Department, who were only too aware of my interest in such matters, had mentioned to me the existence of Cleator's letter in the Liverpool Echo, but I had not seen it myself.

In the meantime, the publication of the letter had produced only a single reply and even that was received after a delay of three days. Fortunately, Cleator's letter has also attracted the attention of the special correspondent of the Daily Express, Mr NE Moore Raymond. He also proceeded to visit P Cleator in order to discuss the proposal that an interplanetary society should be formed.

Following his meeting with Cleator in Wallasey, Moore Raymond had a front-page splash story in the Daily Express concerning Cleator's interests in rocket experiments and his proposal to form an interplanetary society, and it was this article published on September the 20th 1933 that prompted me to write to Cleator on the very same day, inquiring about the Society.

In writing to Cleator, I hasten to add that I had managed to form a small science fiction club (The Universal Science Circle) in Liverpool, but the half a dozen or so members that had comprised the circle, had been disbanded about a year previously due to lack of enthusiasm. I told him that I could still get in touch with members, and I went on to say that one of them had actually been trying some rocket experiments on his own and had given a talk on the subject at a meeting of the Circle.

In the year 1933 the Postal Services were extremely speedy, and I received a reply from Phil Cleator dated for the following day, 21 September 1933. He had received a number of inquiries about the proposed society and rather than deal with them individually in writing, he intended to arrange a meeting when the whole matter could be discussed. In reply to a question in my own letter, he added that he possessed some of the American science fiction magazines, which had been sent to him by Hugo Gernsback, with whom he had been corresponding for some months. He had also been in touch with other members of the American Interplanetary Society.

Cleator's letter of 21 September was typewritten on very imposing looking printed notepaper headed "Scientific Research Syndicate" - Director of Research, PE Cleator, AMIRE, AMIET, and the following phrases were printed on the left-hand margin of the notepaper:

> Chemical physical and electrical research
>
> Analysis and synthetic compounds
>
> Experimental operators manufactured for research work.
>
> Radio and television research
>
> Inventions perfected

A letter dated 25 September 1933 received from Cleator invited me to a meeting to be held at 8 pm on Thursday 28 September at Cleator's home in Oarside Drive, Wallasey and I was asked to bring with me any friends who might be interested in the project under discussion.

About half a dozen enthusiasts turned up that evening and were entertained at Phil Cleator's laboratory, where he showed us an embryo rocket motor and demonstrated the unstable explosive qualities of Fulminate of Mercury. I was so impressed with this demonstration that when Practical Mechanics published a short note on Fulminate of Mercury in its issue of April 1936, I cut out the item, which is still preserved in my press cuttings book. Briefly, it reads as follows.

> "Mercury Fulminate is the mercury salts of an organic chemical known as fulminic acid. It possesses the chemical formula $Hg(O.CH)_2$, which indicates its composition".

After explaining how simply this mercury salt could be made the item goes on:

> "Mercury Fulminate is one of the most dangerous and explosive chemicals known and **on no account whatever should its preparation be attempted**, since, unless special precautions are

> taken, it is liable to explode with extreme violence. Mercury Fulminate is used in the making of cartridge percussion caps, dynamite fuses and for other detonating purposes".

In spite of all the above, I feel quite sure that Cleator knew exactly how to handle this fearsome salt of mercury.

The prospective members enjoyed a pleasant evening at the laboratory and there was no doubt that they were prepared to form a "British Interplanetary Society". Indeed, there seems to have been no discussion whatsoever concerning the name the Society should bear, which seemed to choose itself. Amongst those present were Colin Henry Askham (an erstwhile member of my Universal Science Circle) and Norman Weedall, the latter of whom was to become my best friend for a period of over 45 years. Unbeknown to me at the time, Colin Askham had claims to fame in at least one other sphere of activity.

It became clear at the preliminary meeting of the new Society that "The Big Three" were Cleator, Askham and Johnson. And so it proved. It was agreed that preparations for the first official meeting of the Society should be made by The Big Three, to be preceded by meetings between us at Askham's home (then in Devonfield Road, Orrell Park, near Liverpool) and at Seacombe Ferry (near where Cleator lived) across the River Mersey from Liverpool.

I missed the meeting at Seacombe, which as it turned out, was attended by both Cleator and Askham, and which was held at a very busy public house on the seafront. At the age of 19 years (and like many of my generation at the time) I knew very little about public houses or the consumption of alcohol. So when I peered somewhat tentatively into the busy smoke room and could not immediately spot either Phil or Colin I, shied off and returned home across the river, somewhat disconsolate. As it happened, my absence from the gathering had no permanent ill effects on either the future of myself, or of the Society, as far as I am aware.

Our preliminary meetings soon bore fruit and I duly received a letter from Cleator informing me that there would be a meeting of the Interplanetary Society at 7 pm on Friday 13 October 1933 at the office of HC Binns Esq, Room 15, on the second floor of number 81 Dale Street Liverpool. The Office of HC Binns was typical of a solicitor's or an accountant's office of those days - and indeed of many today - with dark wood and translucent glass panels. Less than a dozen attended this first meeting, but it was immediately resolved that *'The British Interplanetary Society"* should be formed.

Amongst the portentous events of the time, and in connection with the first meeting of the BIS in particular, strangely enough, what stuck in my mind (and rather shocked me at the time) was a rather cynical remark made by a Colin Askham, no doubt the result of bitter experience: *"How long will it be before we all fall out with one another?"*.

From the beginning, it was envisaged that the BIS was not destined to be a local *"Rocket Club"*, but was to be a national - if not an international organisation, and was to be headed by a president. Indeed, had we been content to have remained a small local society, we would undoubtedly have been spared much blood, sweat and tears, and our association would have ended forever at the outbreak of the Second World War - if not before.

The obvious choice for the President was Philip Cleator, the Founder of the Society with Colin Askham as Vice President; I volunteered to be Hon Secretary, while, Miss AC Heaton, MPS became the first Hon Treasurer. Unfortunately, we did not see a great deal of Miss Heaton (the Society's first lady fellow) as the BIS had only been in existence for two or three months when Miss Heaton was to become a Mrs. She was granted an Honorary Fellowship in respect of her interest and her services to the Society and that was the last we ever saw or heard of her. Phil Cleator, himself, took over the additional portfolio of Hon Treasurer as well as that of President.

Other names that spring to mind as having attended the earliest meetings of the Society are those of James Davies, G2OA (a ham amateur radio enthusiast - introduced by Colin Askham), J Toolan, Thomas

McNab, and E Roberts, as well as James A Free Jr. Norman Weedall was confirmed as the first Hon Librarian.

Having appointed the officers, and having arranged for regular meetings of the Society to take place, efforts were made to formulate a legal constitution acceptable to the membership. It was agreed that the Society could only expand as a result of suitable publicity, and as with other "learned " societies (such as we had pretensions of being) we would need to produce an official publication.

In respect to the idea of publicity, there was some continuing help resulting from Cleator's association with Moore Raymond. During his period (about December 1933) and for some days and weeks thereafter, Cleator was featured in an issue of the Daily Express with an interplanetary science fiction serial entitled "Mutiny on the Moon". I remember on one occasion when I posed the question as to how the plot would develop, Cleator smiled blandly and said he hadn't the faintest idea - he just wrote the story from day today.

The officers of the Society duly formed themselves into a temporary Council while the big three proceeded to draft the Constitution. Early publicity was obtained through the pages of the monthly magazine, *Practical Mechanics*, which featured reports on the activities of various scientific societies, and it was to the November 1933 issue of that magazine that I contributed a short report which disclosed that meetings of the BIS were to be held at 81 Dale Street, Liverpool from 6:30 pm until 9 pm every other Friday. Fellowship was available at two guineas (£2-10) per annum, membership at half a guinea (52½p) and juniors at five shillings (25p). The report added rather drolly that all subscriptions may be paid quarterly. All members were to receive free copies of the *Journal* of the Society, which was to appear, four times per year.

Phil Cleator then set a hot pace in offering Honorary Fellowships, somewhat lavishly I thought, and which were sometimes bestowed upon those who obviously merited such an honour and sometimes those who apparently did not. Rather strangely, he used to consider that the acceptance of such Hon Fellowships honoured the Society, rather than the recipients - and I could quite see his point of view.

Notably, Herr Richard and Herr Raymond Thiele were elected as "Honorary Founder Fellows " - "both distinguished scientists, as capable as they're willing to be of service to the Society". But although I was Hon Secretary, I never had any correspondence with them, never knew anything about them, and knew of no services that they might have carried out for the BIS no matter how willing they may have been. I can only assume that they were in some way connected with the German group of rocket experimenters near Berlin, and might possibly have been of assistance to the Society through Phil Cleator at some time or another.

Certainly, one would not begrudge Hon Fellowship when it was obviously appropriate, or to personalities to such as Moore Raymond, Willie Ley (then a leading light in German rocketry) or G Edward Pendray. Cleator names many more in his book, *Rockets Through Space* - "all of whom became Fellows of the Society" - a phrase which I adopted myself and which (as will be seen later) got me into hot water with Ralph Stranger. The idea of names notable in the world of science, all flocking to join the new Society was very gratifying, to say the least. But what was not always made clear was that (whether the honour was theirs or ours) they were being offered and were accepting free Honorary Fellowships. I had an uneasy feeling that whatever may have been the publicity merits of such appointments, the BIS was in danger of acquiring more Honorary Fellows than there were paid up members.

As mentioned earlier in this narrative, not all who had an interest in the "interplanetary idea" also fostered an interest in science fiction, but the ordinary members of the Society in those early days were practically 100% science fiction readers. It was not surprising, therefore, that as Hon Secretary of the Society, and an ardent science fiction fan, apart from obtaining publicity through the reports of meetings that I sent monthly to *Practical Mechanics*, I made a special point of seeking members through the correspondence columns of the American science fiction magazines. The editors of these magazines were sufficiently far-seeing to include the full names and addresses of the readers, whose letters were published in each issue.

Indeed, if the editors of these magazines (*Amazing Stories, Amazing Stories Quarterly, Astounding Stories* and *Wonder Stories*, and its quarterly), following the original example of Hugo Gernsback in *Amazing Stories*, had not included such information in each issue in departments of the magazines such as *The Reader Speaks* and *Brass Tacks*, the development of the science fiction fan activities, and the consequent rise of interplanetary and rocket societies would have taken a lot longer, if they would have arisen at all.

My first venture into the field of science fiction magazine correspondence had taken place early in 1931, when the March 1931 issue of *Wonder Stories* saw a letter printed from one Walter H Gillings (who later was to edit *Tales of Wonder* and *Science-Fantasy* in Britain), while the same month in *Amazing Stories* saw a letter published from one John Russell Fearn. Fearn was destined to become one of the most prolific writers of science fiction stories; both Gillings and Fearn were to become very good friends of mine, solely as a result of the existence of the readers' columns of the science fiction magazines.

There is little doubt, but that story of the British Interplanetary Society, while based in Wallasey in Liverpool between the years 1933, and 1937 was that of a struggle to produce the *Journal* at reasonably regular intervals. From the outset, it was evident that members who were unable to attend the meeting in Liverpool would expect to have their interest maintained by the receipt of a regular Society publication. This would act as a forum for discussion of the interplanetary idea, and would keep members informed of the latest developments leading to the conquest of space. Publicity of various kinds could obtain members, but once they were enrolled, members residing outside the Merseyside area in particular would expect to be offered something more by the Society than the general satisfaction of knowing that their subscriptions were being utilised to further the general aims of the aims and objectives of the Society.

In many ways, this was the fundamental difference between being a local and being a national organisation; and it was going to prove very difficult, with the relatively few paying members that we possessed, to be able to resort to the luxury of a prestigious printed publication, which was what we then had in mind. But one way or another, with the kind of ambitions that we harboured this was a problem that had to be overcome.

The fundamental (if not the only!) difficulty was inevitably that of finance. We did not appear to have the resources to produce a *Journal* at any moment. In the remaining months of 1933 when all possible sources of outside help had been exhausted it fell to Phil Cleator, himself, not only to edit and compile the *Journal*, but to illustrate it - and also to pay for it!

Cleator had very generously announced that he would put up a price of one guinea (£1.05) to be won by the member who submitted the best black and white illustration to be used as the cover design for the forthcoming first issue to be published of the *Journal*. The names of the artists were not to be disclosed before the entries were adjudicated, but at the meeting at which the winning design was chosen by the members, it was revealed that Phil Cleator had saved himself, a certain amount of additional expense by winning his own prize.

Summary of events The year - 1933

- My sources of information
- Earlier *"Interplanetary Societies"*
- The American Interplanetary Society
- The American Interplanetary Society and science fiction
- Science, science fiction and interplanetary travel
- Philip Ellaby Cleator *Chambers' Journal*
- PE Cleator's article in, January 1933
- The predictions of W A Conrad The Liverpool Echo, August 1933
- P. E Creator's letter to the Liverpool Echo, 8 September 1933
- N E Moore Raymond of the Daily Express is interested
- Moore Raymond's article in The Daily Express, 20 September 1933
- LJ Johnson writes to P Cleator
- *"Universal Science Circle"*
- P E Cleator and the science fiction magazines
- The meeting at Cleator's laboratory, 28 September 1933
- The explosive qualities of Fulminate of Mercury
- The British Interplanetary Society gets its name
- Colin Henry Askham, and Percival Normal Weedall
- A meeting missed at Seacombe Ferry
- The first official meeting of the British Interplanetary Society. Friday 13 October 1933
- The British Interplanetary Society is founded; officers are appointed and a Constitution is to be drafted
- *"When will we all fall out with one another"?* CH Askham
- Early members of the Society
- The Constitution and the need for publicity
- *Practical Mechanics* November 1933 gives details of BIS membership
- *Mutiny on the Moon* by PE Cleator, December 1933
- Herr Richard and Herr Raymond Thiele Honorary Founder Fellows
- Too many Honorary Fellows?

- Seeking publicity for the new Society
- The value of readers' addresses in the American science fiction magazines
- Walter H Gillings and John Russell Fearn, March 1931
- The necessity for a regular Society publication.
- The question of the cost
- P E Cleator edits, composes and pays for the *Journal*
- P E Cleator, as an artist, wins his own prize.

THE YEAR 1934

The first issue of the *Journal* duly appeared dated January 1934. Cleator's cover design was somewhat reminiscent of artist Frank R Paul of Gernsback and science fiction fame. It depicted a black rocket-liner taking off against a cubical background of skyscrapers, the night sky showing a huge white full Moon, with a sprinkling of stars as an accompaniment. This scene was to adorn the cover of the first four issues of the *Journal* of the Society, all of which appeared during the year 1934.

The first issue of the *Journal* was very sparse in appearance, and in content - as might have been expected - octavo sized, six printed pages, of which the cover illustration took the first, Cleator's editorial, "Retrospect and Prospect" took the best part of pages 2, 3 and 4, page 5 gave details of membership, while page 6 contained a (free) full-page advertisement for *Chambers' Journal*, with particular reference to Phil Cleator's article, "The Possibilities of Interplanetary Travel", which, of course had been published in the magazine about a year earlier.

Copies of the rather skeletal *Journal* were circulated not only to the relatively few members of the Society (then consisting of about 15 paying members) but were distributed gratuitously to newspapers, journals, and various eminent personalities - not to mention the numerous Honorary Fellows. The copies of the *Journal*, perhaps somewhat surprisingly, did not fall altogether upon stony ground and the result was a gratifyingly increase in membership.

It might be of interest, at this stage, to refer to a letter headlined "Misspelt Credit", written by PE Cleator and published in the September 1951 issue of the *Journal* of the British Interplanetary Society. Apparently, in the July of 1951 issue of the same publication, it had been reported during the course of discussion following Professor Haldane's lecture to the Society on "Biological problems of spaceflight", Arthur Clarke had mentioned that one of the Society's earliest members, Mr Colin Askham, had actually paid for the first issue of the *Journal* to be printed. However, as so rightly claimed by Cleator in his letter to the *Journal*, the credit had indeed been misplaced, as the cost of producing the first issue had of course been met by Cleator himself.

The truth of the matter was, as Cleator explained in this letter to the *Journal* of the Society that Vice President Colin Askham had been able to make an approach to John Moores, the Littlewoods Pools magnate, who had agreed to the second and third issues of the *Journal* being printed free of charge at Moore's printing establishment in Brownlow Street Liverpool.

For reasons of his own, John Mores did not wish to receive any public acknowledgement of his assistance, which as far as I know, received no publicity until Phil Cleator's letter appeared in the September 1951 issue of the *Journal* of the Society.

Even assistance such as this welcome though it may have been, only postponed what seemed would be the inevitable end, when unless the membership numbers showed an early substantial increase the Society would become moribund - and that would be the end of it as a national institution.

In the meantime, I was producing monthly mimeographed sheets which were issued to members of all grades, giving general information, news, and the minutes of the meetings of the Society. However, valuable time had been gained by the gift of two issues of the *Journal* to the Society by Mr John Moores.

As indicated by the report in *Practical Mechanics*, meetings of the Society, were held from 6:30 pm until 9 pm every other Friday; 6:30 pm was in fact, an extremely early hour at which to start a meeting, as was 9 pm at which to finish. It was also found to be inconvenient for meetings to continue to be held at Binn's office on Dale Street. That office had to be closed and the meetings terminated by 9 pm was not without its advantages. Cleator and Binns were devotees of all-in wrestling and were able to make their way over the short distance to Liverpool Stadium in order to witness one of their favourite sports.

Not long ago (in 1978) I read in the pages of the Liverpool Echo, that the drinking licence having been renewed, after suspension for various reasons, all in wrestling would be resumed at Liverpool Stadium (still!) on Friday evenings. Things have changed little in over 40 years.

Efforts were being made to find an alternative meeting place for the Society, especially as the members in attendance were showing gratifying increases. Indeed, at the meeting held on 6 September 1934, the attendance record (such as it was) had been broken. So many persons were present that the room became stuffy and sufficient chairs could not be found to accommodate everyone. At this meeting, the possibility of the formation of the London branch had first been raised, and it was resolved that such a branch would be established just as soon as the numbers of members living in the London area justified this course of action.

The meeting of 5 October 1934 was held at my home address 46 Mill Lane commencing at 6:30 pm. A subsequent meeting was held (this time commencing at 7 pm) at the Prince Cafe on the corner of Old Hall Street and Tithebarn Street, not far from the Liverpool Town Hall. No time was specified by which the meeting should end. Towards the end of 1934 meetings were to commence at 7:45 pm and would be held at the McGhie's Cafe 56 Whitechapel; this location was to afford a regular meeting place until the Society was finally transferred to London in February 1937.

McGhie's Cafe was more elegantly known as The Hamilton Cafe, which introduces a rather odd little coincidence: my older daughter, Valerie, was born on 13 October, the date on which the BIS held its first official meeting - in 1941 - and in 1969 she married James Hamilton.

Fortuitously the demolition squad has so far spared No. 56 Whitechapel, in Liverpool, although the premises are poised perilously near the end of the line - an honour reserved for the neighbouring building, occupied by the National Westminster Bank. At street level No. 56 is currently tenanted by "Tape Electronics", where I regularly buy earpieces for my transistor radio and on the first floor, at the top of the wide staircase up which trod pioneering members of the BIS in the middle Thirties, is Sukies Unisex Hair Style, which I do not frequent.

The first of my letters to the American science fiction magazine, asking for members for the BIS was to appear in the April 1934 issue of *Amazing Stories*, followed by a similar letter in the November 1934 issue of *Astounding Stories*. Although current issues of these magazines (dated two months ahead of their actual publication dates) were usually available from a limited number of bookstores in Britain at prices ranging from 1-3d (6p) to 1-6d (7½p). The vast majority of science fiction readers of the time were introduced to the cult by picking up so-called "remainder" magazines. These could be obtained from branches of Woolworth's and from second-hand book shops at 3d (1p) per copy.

"Remainder" magazines were unsold copies sent over from America as ships' ballast, and sometimes had a special mark rubber-stamped on to the front cover. Other copies, had the top right-hand corner of the magazine cut off; still others had had front covers ripped off and sent back to the publishers in order to obtain credit refunds. The unsold copies took some months to percolate through to the British market, usually, arriving some 6 to 12 months after publication in the US.

One of the earliest results, obviously, following from my letter that had appeared in the April 1934 issue of *Amazing Stories*, was a letter received from a 16-year-old lad named Arthur Clarke of Bishop's Lydeard in Somerset. His letter was dated 30 July 1934; on 25 August 1934, an inquiry was received from one Eric Frank Russell (who said he had seen my letter published by Teck Publications -*Amazing Stories*), and the 14 September, the same year, a letter arrived from Walter H Gillings, with whom of course, I had previously corresponded during 1931 with a view to forming science fiction clubs throughout the country.

The letters in my file received from Arthur Clarke and dated before December 1934 are typewritten copies I made of his handwritten letters; it appears that I had an arrangement at the time with Phil Cleator that I would copy out any letters received on behalf of the BIS and would send the originals to him. I also sent Cleator carbon copies of my replies, so that he was kept informed of what I was doing, within the Society. Obviously, such a system involved me in a lot of extra work, and towards the end of 1934 Cleator and myself merely exchanged carbon copies of our respective replies to any enquiries received.

Arthur Clarke's letter, read as follows.

Dear Sir,

Please could you send me particulars about your Society, as I would very much like to join it. I am extremely interested in the whole subject of interplanetary communication, and have made experiments with rockets. I am 16, have extensive knowledge of physics and chemistry and possess a small laboratory and apparatus with which I can do some experiments in this line. I am also interested in astronomy, and have some small homemade telescopes. Although I am afraid I could not attend your meeting in Liverpool, I would like to read your *Journal*. I understand the problems that have to be faced before even a flight of a few hundred kilometres can be made by rockets, as I have read David Lesser's book, *The Conquest of Space* and the various American magazines (*Wonder Stories* etc.) which contain stories of this subject.

I see your President, Mr Cleator, wrote a story, *Martian Madness* in *Wonder Stories*. I thought it was a very good little yarn. The enclosed cutting from the Daily Mail may interest you. It annoyed me intensely as it is chiefly a concoction of lies calculated to retard progress in rocket research which is not the intensely national and militaristic thing this Frenchman thinks it is. I enclose an envelope for your reply which I shall eagerly await.

Yours Sincerely,

(Signed) A.C. Clarke

The letter received from 'E F Russell, then resident in Bootle, near Liverpool, read as follows:

Dear Mr Johnson.

I was very much interested in your letter anent The British Interplanetary Society given publicity by Teck Publications Inc. of America. Enclosed is the first instalment of a seven part serial of mine, that is now appearing in the *Ida and Victoria* magazine a private organ of Messrs Fred Braby & Co. Ltd. of Liverpool and London. The serial is intended to deal exhaustively with this subject and it will incorporate details of the activities of interested parties throughout the world. I shall be pleased to give such information about the BIS as is of interest to the layman.

If like the American IS and the Cosmos Science Club, you publish a journal and would like to reprint this article you can probably get permission and the loan of blocks for illustrations from the address attached.

Most sincerely yours,

(Signed) EF Russell.

Mr Russell, then went on to give the address of the editor, A.M.I. Baker, Esq., c/o Messrs. Fredk Baby and Co. Ltd. Ida Works, Deptford S.E. 8 (Until he retired, early at the age of 55, in order to devote his time to science fiction writing. Eric Frank Russell was a commercial traveller with Messrs Frederic Brady,& Co. Ltd., manufacturers).

Walter H Gillings wrote as follows:

Dear Mr Johnson.

It is now two years since I corresponded with you, as secretary of the Ilford Science Literary Society (now, alas! long defunct) with regard to the possible publication of a magazine for the many science fiction lovers that now exist in England. Having found your new address (and new position - congratulations) in the January (1934), *Amazing Stories*, which I have only just secured, I hasten to write to you to reopen the discussion. I am as keen as ever, on this subject, which I fully explained to you over the period, beginning 21 May, and ending 12 August 1932. So, I will not go over the ground again. I will, however, ask you to be good enough to write to me and say whether your present members are interested in the idea, and whether they would be prepared to give the magazine their support.

I was thinking that, if all the organisations of science fiction fans in the country - and there must be quite a few now! got together and did their little bit towards financing a magazine, it would be the most satisfactory way of doing it, and doing it fairly cheaply, too. For I am still willing - nay, anxious - to do the editorial part of the work myself, providing I was not burdened with too much expense.

I am going to write to *Amazing Stories* on the subject, and see if they will publish my letter. I wrote to *Wonder Stories* about it some time ago, but nothing appeared. They probably thought such a mag would be a serious competitor!

However, I should very much like to have your views, and how you are getting on, after all this time. I see your friend, Mr Fearn is reaping a few laurels in *Amazing Stories*. All power to his elbow! Many's the shot I've had at doing likewise, but I don't seem to get the time to finish anything that I start. For I'm still an energetic newshound, of course. I'd find the time for a magazine, though, if I could produce one that would be certain to be a success. And it should, with all the enthusiastic readers of science fiction we have here these days. But in spite of this, I am still unable to persuade a publisher to cater for them.

Did you see *Scoops*, "Britain's only science fiction weekly"? There was a noble experiment that went flop because it was not properly handled. Catering for schoolboys! Huh! I was writing a story for them, though, and had nearly finished it, when the poor thing gave up the ghost!

Still, I shall be very pleased to hear from you - and by the way, do you remember the magazines you so kindly obtained for me in Liverpool? I wonder if you could again oblige me by enquiring of your fellow members whether they can supply me with any of these magazine numbers (followed a list of Science Fiction magazines that he required including issues of *Amazing Stories, Wonder Stories, Amazing Stories Quarterly, Wonder Stories Quarterly* and *Strange Tales*.

I shall be eternally grateful if you can secure some of these elusive numbers for me in your part of the world.

Eagerly anticipating your reply.

Yours very sincerely

Walter H. Gillings

PS I should be glad also, if you could spare a specimen copy of your Society's journal.

In fact, the address to which Gillings had sent this letter was that of P Cleator in Wallasey, who passed it on to me at 46 Mill Lane, Old Swan, Liverpool 13, which was the same address at which Gillings had written to me in 1931 and 1932. And the letter from me that he says was published in the January (1934) issue of *Amazing Stories* must have been the one that appeared in the April 1934 issue, as I am sure that no such letter appeared in the January 1934 issue. During this period, about which I'm writing and despite Gillings' reference in this letter to the "many science fiction readers in England", science fiction enthusiasts and advocates of interplanetary travel were generally very isolated souls.

And it became my very pleasant task as Hon Secretary of the BIS and therefore, the hub of the wheel of science fiction and interplanetary correspondence to accumulate the names and addresses of devotees to both interests (most of whom existed unknown to one another) and to put them into touch with each other.

Many of the names that I have in mind would now form the foundation of a "Who's who in science fiction" and their associations with each other came more and more into prominence as the BIS grew in numbers. This culminated in the formation of a London branch of the Society, the rise of the Science Fiction Association, which was founded in Leeds in January 1937, and the transfer of the Headquarters of the BIS to London in February 1937.

To revert to Walter Gillings - Walter was a kind of extreme science fiction fan who while retaining a general interest in the idea of interplanetary travel, was mainly interested in editing and publishing a science fiction magazine in Britain. In this aim, of course, he eventually achieved success.

It was over two years after his letter to me in September 1934, in fact, in November 1936, that Walter finally joined the BIS, and then it was mainly because of the formation of the London branch of the Society, consisting as it did of 99% of science fiction fans, and because of the imminent transfer of the headquarters of the BIS to London, almost as though in spite of himself. He was destined to play an important part in the future development of the Society, even extending into the immediate post-war years.

Almost inevitably, enquiries regarding membership of the BIS progressed from official correspondence between prospective members, and the Hon Secretary to personal letters between "Les", "Eric", "Wally" and so on, more because of a general interest in science fiction than because of enthusiasm for interplanetary travel.

In October 1934, the first of a new series of quarto-sized Mimeographed *Bulletins* was issued, consisting of ten pages.

This new *Bulletin* was produced by James Free Jr. and myself, in order to replace the monthly duplicated sheets giving reports and meetings of the Society, and further to sustain the interest of the members.

Such care and effort went into the production of these issues of the *Bulletin* that the quarto sheets appeared in two columns of typewriting with even edges on both sides of the column.

The new *Bulletin* started off with an editorial by Leslie Johnson, entitled "Our First Anniversary". It might prove interesting if I reproduce it here.

Saturday, 13 October marks the conclusion of the first year of existence of the British Interplanetary Society, for whom on the same day of 1933 was held the official inaugural meeting of the Society. This was held as a result of an informal discussion at the home of P.E. Cleator who was later elected President.

Friday the 12 October was fixed for celebration to round off the first year's work, but due to unforeseen circumstances, and to the feeling that it really was not appropriate to celebrate before the time of the actual anniversary itself, this function was postponed. It will be held more towards the end of the month or in early November.

During 1933-1934, the Society has been governed by depleted Council, due to certain of the original members finding it impossible to fulfil their various offices. Nevertheless, despite this fact, it is possible to report the conclusion of a year of remarkable progress, surpassing all our expectations. The election of the new Council will occupy the attention of the members in the near future. Meanwhile, there are certain changes being discussed, which could result in even more progress in the coming years.

It is easy to trace the rapid rise of the Society, directly to the publication of the *Journal*. The publicity resulting from this has been instrumental in attracting to the Society those members of the general public, who could already foresee the eventual conquest of space. To the others the existence of this Society must have come as a surprise but there is little doubt that many of these have been won over to the course of the promulgation of the interplanetary idea.

There still remains, nevertheless, a vast majority, who steadfastly refuse to discuss seriously the possibility of space travel, and still less would they contemplate association with a Society, such as ours. It is this class of person that we would educate to the realisation that condemnation is no argument. Let them study the matter from a scientific standpoint, and let them draw their own conclusions, not from the thoughtless prejudice, but as a result of a close study of the factors involved.

In the dissemination of knowledge concerning the problems which await solution before the accomplishment of actual interplanetary travel, the *Journal* is a vital factor. The present quarterly production is, however, a very unsatisfactory arrangement, and to supplement this, the *Bulletin* was brought out. This was nevertheless, mainly of interest to the members, and was of little moment to those outside the Society, even if they ever had the opportunity of perusing a copy.

In this connection, the new *Bulletin* is being reproduced as an experiment by the secretarial staff, and we venture to predict that it will produce a very favourable response. The opinion of members of this innovation will be welcomed by the editors.

Our Membership

A gratifying and steady increase in the membership has been in the feature of the past year. This increase has just taken the form of an addition of 15 members every quarter, which is just over one per week. What it may seem to lack in quantity, moreover, it will make up for in quality. The Council, however, are very satisfied with this result of Society propaganda and are hoping that this steady influx of new members will be continued in the future.

A visit to Germany - its results

An event of the greatest importance, was the visit by Mr Cleator to the German experimenters last January. Through Herr Willy Ley, Vice President and secretary of the E.V. Fortschrittliche Verkehrstechnik, whom he met in Berlin, we have established communication with the societies in America, Austria, France, and the USSR. The Society is now in close touch with the officials of the various bodies in these countries, and the friendliest relations exist between ourselves and the foreign enthusiasts.

Through Herr Ley, also, there was obtained a list of those in England who had written for particulars of the German Society. Amongst others on this list was the name of Professor A Low and this famous scientist unhesitatingly associated himself with the work of the British Interplanetary Society. As well as Professor Low, we enrolled the famous experimenter, Mr H Grindell Matthews, whose work on lethal rays in the past was once the subject of much comment. Mr Matthews has taken up rocket experiments in the laboratory which he is building especially for this purpose.

Radio research has been carried out by the radio section of the Society under the Vice President, Mr C Askham. At the moment, Mr Askham (G6TT) and Mr J Davies (G20A) are busy on the construction of new and improved transmitters with which they will explore the region of the ultrashort waves.

With this line of research, and the promised experiments of Mr Matthews, the Society foresees a future as laden with the greatest possibilities, and regards past achievements with the profoundest sense of satisfaction.

In facing the coming year, we have at least the knowledge that the Society is now established on firm basis. We have still many problems to face, but, in arriving at their solutions, let us hope that we can solve them as we should - as a Society, determined to further the promulgation of those ideas, which, sooner or later will result in the conquest of space. - Editor

---

Cleator's visitations to and associations with, Mr H Grindell Matthews, were (no doubt rightly, at the time) something of a vague mystery to the various members of the Council of the Society. Cleator kept hinting at strange portents and possibilities, and we have to refer to Cleator's article, "Matters of no Moment" in the March 1950 issue of the *Journal* of the British Interplanetary Society (16 years later !) before any light seems to have been thrown on the affair.

According to Cleator, H Grindell Matthews had been very busy with a scheme for a massed rocket barrage against fleets of hostile bombers, and had provided himself with an extensive mountain-top research station on Tor Cloud in the vicinity of Swansea, surrounded by a high electrified fence, concrete observation trenches and even having his own private airfield! Cleator added that Matthews was convinced of two things - that a World War was on the way, and that rockets would play no small part if it came.

Cleator, who was not interested in the rocket as a weapon of war, refused an offer to take up residence at Tor Cloud and work there with Matthews Neither, it seemed was the War Office interested (strangely enough), and having encountered one frustration after another, Matthews died in September 1941, without having seen his scheme put into operation. That it was ultimately used, Cleator stated, was evidenced by anti-aircraft batteries called "Z" guns which appeared a year or so after the death of Matthews.

The *Bulletin* dated October 1934 also contained a note from Herr Ing. Guido von Pirquet (one of the earliest advocates of construction of a space station) to the effect that the majority of the books listed as German by Herr Ley in No. 3 of the *Journal*, were in fact of Austrian in origin. I had added a note to the effect that the main point of interest was that German was the language in which they were written.

Colin Askham attributed an article entitled "Radio in Interplanetary Communication" in which he stated that they were working on the assumption that communication between planets and spaceships would be on ultra-high (quasi- optical) frequencies. I added a note to the effect that absorption of any wide band of frequencies by the heavy-side layer seemed hardly conceivable. Should any part of the wave band be found to be absorbed, it would only mean that some other frequency would have to be used. I went on to say that echoes of radio broadcasts had been received several seconds after transmission, indicating that a distance had been travelled of about one million miles. Should this be accepted as actual fact, there would be no doubt regarding the possibility of interplanetary radio communication.

In re-reading the preceding paragraph, it seemed to strike a familiar note: in recent years, there have been articles in *Spaceflight*, a publication of the British Interplanetary Society, discussing the possible significance of long delayed radio echoes, which it was suggested might be typical of an attempt at communication by an alien interstellar probe! The name of Duncan Lunan of Glasgow University has been associated with these investigations, which are fully discussed in his book, *Man and the Stars - Contact and Communication with Other Intelligences*.

Also included in the October 1934 issue of the *Bulletin* was quite a long letter from J Strong, continuing his plea that the name of the British Interplanetary Society should be changed to some more "practical" designation such as "The British Rocket Society". Said Mr Strong: "Is the Society to stick purely and simply to the problem of interplanetary travel (in which case we shall be a long time getting anywhere) or are we to embrace the many parallel developments which will forerun the main issue?"

In April 1934, the American Interplanetary Society had become "more practical" and by changing its name to "The American Rocket Society" - but "If you the change of the name, it's not the same" - and a glance at the issues of *Astronautics* (the publication of the American Rocket Society at that time) would indicate that they seem to have been a greater interest in developing rockets, for their own sake than in interplanetary travel.

James Godwin Strong FBIS, was still a well-known member of the British Interplanetary Society 40 years later and the author of books on interplanetary and even on interstellar travel, including a book entitled *Flight to the Stars*. I often wonder whether he ever changed his views concerning the name of the Society. Even today, the issue is still an occasional bone of contention in the correspondence columns of *Spaceflight*, the writers in these later days probably little realising that this was a bone that was well chewed over as long ago as 1934.

The *Bulletin* reported on the meetings of the Society held on Friday, 5 October and on Friday, 19 October 1934. In his report on the first of these two meetings, the Hon Secretary revealed that the members present had unanimously voted to retain the original name of the Society, and that all members who had written to him following the receipt of Mr Strong's letter and express the same view.

Ing. Guido von Pirquet had joined the Society's "Information Bureau", and on receipt by him of an International Reply Coupon, he was prepared to solve any mathematical problem, that might be submitted by the members.

Mr Ernst Lobell, technical adviser of the Cleveland Rocket Society, had written accepting Honorary Fellowship of the BIS, and would represent the Society in the Cleveland group.

Mr James Free Jr. was appointed Assistant Hon Secretary, as there had been a steady increase in the secretarial work; there were discussions about making the Society notepaper available to members, and about designing a BIS lapel button (which never eventuated, at least under the Liverpool regime of the Society).

A special members' notepaper with a design by Mr K Chapman, was shown to those present; the heading showed a disc, about as a big as an old Crown Piece, with (shades of 20 July 1969!) a spaceship taking off from the moon, making its way towards the "Full Earth" shown in the background. This was depicted in black and white, with "The British Interplanetary Society" inscribed on a rather flowery scroll (Victorian style) around the central design. In later years, this could almost have been identified as a scene from Apollo Moon landing.

At the meeting of 5 October 1934, an editorial committee was set up to supervise the production of the *Journal*, this committee consisted of Messrs. Cleator, Free and Johnson.

At the meeting held on 19 October Mr Free was further appointed to the post of Hon Treasurer, which since the resignation of the original Hon Treasurer had been held temporarily by Mr Cleator.

The Council of the Society, having become depleted in numbers, was brought up to its full complement of five members all founder fellows of the Society. These consisted of the President, Mr P Cleator, Vice President, Mr C Askham, Hon Secretary, Mr L Johnson, the legal adviser, Mr H Binns and Radio Technician, Mr J Davis. Mr J Free Jr. Hon Treasurer and Assistant Hon Secretary would become a sixth (ex-officio)member of the Council.

The second issue of the new series of the *Bulletin* was dated November 1934, to which I contributed, an editorial "Scientifiction, and Interplanetary Travel"; there was also "The Astronaut", (a poem) by E

Russell. The issue also included reports on meetings of the Society, held on Friday, 2 November and Friday 16th November 1934.

At the meeting held on Friday 2 November 1934, the chair was at first taken by the President, Mr P Cleator, but as he was compelled to leave before the end of the proceeding, Mr T Ashcroft (member-Liverpool) took the chair for the remainder of the meeting. (Amongst the members present, there was a suspicion that there might have been a particularly attractive bill of fare on that night of the Liverpool Stadium)

The necessity of having an Hon Librarian was highlighted by the presentation to the Society of copies of the various "scienctifiction" magazines by Mr F Knight (member - Walsall), as well as by the Vice President and the Hon Secretary; Mr Norman Weedall, (Founder Fellow - Litherland) kindly consented to fulfil this office.

A Dublin member had written to suggest that a Dublin newspaper might be approached with a view to the publication of an article on interplanetary travel; on the other hand, a proposal had been made by Mr E Russell (member -Bootle) that efforts should be made to obtain a broadcast about the Society by the BBC. It was reported that the notice about the Society had appeared in *Astounding Stories*, one of the American Scientifiction magazines. A free discussion, lasting an hour and a half, took place on the subject of the temperature of an object in space

At the meeting which took place on 16 November, the chair was again taken by Mr Cleator and members were informed that as decided at the recent Council meeting, the *Bulletin* of November 1934 would be produced in the new (large) size. Publication of future issues would depend upon the of course to be followed in the coming year with regard to both the *Journal* of the Society and the *Bulletin*. A decision would be made by the Council in the near future.

A free discussion was held on the subject, "Can we plot a path for a spaceship from Earth to one of the planets?" suggested by Mr C Askham (Vice President and Radio Director of the Society).

The new series of quarto-sized *Bulletins* were at least a courageous attempt to interest, and to retain those members that the Society already possessed, but it was realised that it might become more difficult to inspire new members to join the Society, if the *Bulletin* in fact completely replaced the *Journal*.

By the end of 1934, however, it had become clear that the problem of the *Journal* was one that had to be faced squarely by the Council of the Society. Indeed, the year 1934 had seen the unfortunate development of that situation foreseen by Colin Askham at the first meeting of the Society, when he had the prophesied difficulties that might arise in personal relationships.

The issues of the *Bulletin* for October and November 1934 contained hints that all was not well in the management of the BIS. For instance, the future form of the *Journal* itself was uncertain; the address of the Society was given as " 46 Mill lane, Old Swan, Liverpool 13 " (my home address) instead of Oarside Drive (Cleator's address) and the informal celebration that was to have been held on the occasion of the fifth anniversary of the foundation of the Society, never took place.

It had become clear that dissension existed between Mr P. Cleator and those who supported other points of view concerning the manner in which the Society should progress. In the early months of the Society's existence, when it was possessed relatively few members, (some of these having already previously been personal friends of the president), it was understandable that things should have been allowed to go Cleator's way, bearing in mind the natural diffidence that existed towards the Founder of the Society.

However, as time went by, more and more individuals of ability and independence (such as Eric Frank Russell) were joining the ranks, and as a result, the voting power was gradually shifting away from Cleator. At first, this became apparent at the ordinary meetings of the Society, then later at the Council meetings. The speed of this transitional process, as far as the Council was concerned, depended either upon natural wastage, or the manner in which personnel on the Council were replaced at the Annual General Meeting.

The battle for power was very evident to those who could read between the lines in the pages of both the *Journal* and the *Bulletin* of the Society published during this period, as well as in the pages of the *New Columbus* (about which more later).

How this battle was progressing from time to time, was epitomised in the manner in which the Society's Headquarters address was shown in the publication of the Society and may be summarised as follows:

The *Journal*, January 1934 - Edited by P Cleator, the Secretary, 34 Oarside Dr, Wallasey, Cheshire (P Cleator's home address)

The *Journal*, May 1934 - Edited by P. Cleator (address as above)

The *Journal*, July 1934 Edited by P Cleator; The President at his home address. The Secretary at his home address 46 Mill Lane, Old Swan, Liverpool 13.

The *Journal*, October 1934 - Edited by P Cleator; Headquarters, Cleator's home address. Hon Secretary-Treasurer. 46 Mill Lane.

The *Bulletin*, October 1934 - Edited by L Johnson. The Hon Secretary at his home address;

The *Bulletin*, November 1934 - Edited by L Johnson - a special note stating that the address of the Society was "now" 46 Mill Lane

The *New Columbus*, January 1935 - Edited by L Johnson and E Russell; The Hon Secretary 34 Oarside Drive, adding - items re The *New Columbus* to 46 Mill Lane!

The *Journal*, May 1935 - Edited by P Cleator - The Secretary, 34 Oarside Drive

The *Journal*, October 1935, edited by P Cleator - astonishingly no address given at all, to which requires to be made concerning the Society!

> The *Journal*, February 1936 - Edited by P Cleator. The Secretary, 46 Mill Lane;
>
> The *Journal*, June 1936 - Edited by L Johnson, special emphasis placed upon the point that the official address of the Society was 46 Mill Lane, and that all enquiries should be made to the Hon Secretary
>
> The *Journal*, February 1937 - Edited by L Johnson, all communications should be addressed to the Hon General Secretary at 46 Mill lane, or to the Hon Secretary of the London branch at 95 Forest Road, Walthamstow, London E. 17

At this distance in time over 40 years later, all of the above convolutions must seem to be very bizarre if not altogether hilarious (or even unnecessary!) but at the time, they represented a grim reality of life to those who in their various ways had the long-term interests of the BIS at heart. And what any members living outside the Merseyside area may have thought of all the gyrations, I shudder to think.

However, there was inevitably much good reason behind the battle of the addresses. Mr P Cleator wished to be the prime recipient of the Society's correspondence and thus to be in a position to pass on routine queries to the Hon Secretary. The situation might have seemed to be reasonable and acceptable up to a point, but unfortunately Cleator did not always appear to be very communicative. He knew what everybody else knew and what everybody else was doing, but there were many occasions when the rest of Society had no idea about the activities in which he may have become engaged (no matter how meritoriously) in the name of the BIS.

Even more puzzling in its way was the manner in which Cleator seemed unable to distinguish between "The Secretary" and "The Hon Secretary". This attitude persisted even as far as his article in the *Journal* dated March 1950. It is difficult to accept that a person of Cleator's undoubted erudition could fail to see the implications of the two descriptions relating to the office.

The major bone of contention was, of course, the question of the Society's *Journal*, all issues of which from January 1934 to February 1936 inclusive had been edited by P Cleator. I, myself had the task of editing the last two to have emanated from Liverpool, and which were dated respectively June 1936 and February 1937. The position was, indeed, further complicated by Cleator's very generous monetary contributions towards its continuation - and he was the most competent candidate for the post in any event...

Apart from the contributions by Cleator and John Moores, Colin Askham had made a loan to the Society of the amount of £3, being the amount of the deposit that was paid in order to enable the Society to purchase a duplicating machine. The £3 was to have been repaid to Askham when circumstances permitted, but when a year had elapsed, and the amount had not been repaid, Askham stated that the amount could be put into the Research Fund, whenever we found ourselves in a position to do so. We did, however, during the course of the year, find the sum of £9, this being the balance of the purchase price of the duplicator.

I, myself, had bought a typewriter, which was used mainly for Society business and while Eric Russell was quite adamant that the Society should have paid for this, as well as for the duplicator, the Council refused to do so - and obviously they could hardly have been able to find the funds to do so, in any event.

The year 1934 was, of course, notable for the admission of new members in the form of Arthur Clarke and Eric Russell to the Society, as well as the return of Water Gillings, as one of my correspondents. The correspondence of Gillings was inevitably on the subject of the possibilities, not of interplanetary travel, but of the publication of a British science fiction magazine.

Having paid his 5s.0d (25p), this being the annual subscription for the associate member (in view of his youth) Arthur Clarke, then became one of my liveliest correspondents describing, amongst other things, how he had fitted and dart like wings to penny rockets and launched them off a roof. In a similar manner to what I had accomplished myself, he had a homemade telescope, consisting of a one-and-a-half-inch achromatic lens, a focal length of 20 inches, and the lens of one-inch focal length as an eyepiece. With his telescope mounted on a Meccano tripod, he had been able to make numerous drawings of the Moon's craters.

Lectures given at meetings of the Society were reproduced and circulated to members and accordingly, Arthur undertook to compose a lecture on "Life on Mars", this planet being one of his favourite subjects, even then. Unfortunately, Arthur's lecture on the planet Mars was lost in the post and was never found Having written it in his own fair hand (such as it was), he had retained no copy and the effort was wasted.

Arthur did not have much luck with his early efforts on behalf of the Society. Having lost his manuscript of the lecture on Mars, undaunted, he got work to again and produced an article, "The problem of Mars", intended for publication in the second issue of the *New Columbus*- a second issue - which never appeared.

In the meantime, he thought he would avail himself of the facilities of the Society's "Information Bureau" and apart from posing a problem on radio for Colin Askham, he wrote to me - "The velocity of light in any substance depends up on the refractive index. Sodium has a refractive index of 0.19, so (contrary to Einstein's theory) the velocity of light in a sodium film comes to about 1,000,000 miles per second. How do you explain this?"

With the superior knowledge of a Hon Secretary of an interplanetary society and the maturity of a 20-year-old, I assured Arthur (with a diagram to reinforce my deposition) that the velocity of light passing through any medium is always less than 186,000 miles per second - or certainly not more. This is because (as I explained) the ratio of the sine of the angle of incidence to the sine of the angle of refraction is always constant. He did not come back further to dispute this issue, and on re-reading my apparently very learned reply to Arthur's question, I'm staggered to think how much I must have forgotten in the meantime.

Arthur went on to say that he could not afford to pay the 2s.6d. (12½p) per hundred sheets for the special members' notepaper, as he undertook a very great volume of correspondence each week, and the number of sheets he would require would prove to be prohibitive.

Whoever his other correspondents may have been, by the end of 1934 they included Walter H. Gillings, who had supplied Arthur with some of the science fiction magazines he had been seeking.

Following the receipt of the first letter to me from Eric Russell, and a personal visit to my mother's shop in Mill lane by all six foot three of Eric, correspondence commenced between us although we lived only about 10 miles apart. At first, this was centred mainly around the possibility of Eric's articles on "Interplanetary communications" appearing in a publication of the BIS, but rapidly developed into regular exchange visits to each other's homes and collaboration in numerous projects involving Science Fiction and Interplanetary Travel.

Summary of Events : The Year 1934

- The first issue of the *Journal* appears dated January 1934
- Contents of the first issue of the *Journal*
- The *Journal* distributed
- The problem of future issues of the *Journal*
- Askham approaches John Moores re the *Journal*
- Moores seeks no publicity for his assistance in printing two issues of the *Journal*
- Valuable time gained as a result
- Details of meetings of the Society
- Cleator and Binns attended all-in wrestling at Liverpool Stadium
- All in Wrestling renewed at Liverpool Stadium (1978)
- Efforts made to find an alternative meeting place
- The possibility of the formation of a London branch
- Alternative meeting places
- Later times arranged at which to hold meetings
- The Hamilton cafe, 13 October (and Valerie Hamilton)
- Number 56 Whitechapel, Liverpool, today
- Letters written by L Johnson appear in the American science fiction magazines Clarke, Russell, and Johnson respond to letters printed in *Amazing Stories*
- Original letters received plus copies of replies sent to Cleator
- Arthur Clarke's first letter to the BIS
- Eric Frank Russell's first letter to the BIS
- Walter H Gillings writes again on 14 September 1934
- The splendid isolation of science fiction and interplanetary devotees
- The BIS attracts science fiction names of the future
- Walter Gillings and the BIS
- Enquirers evolved into personal friends
- The *Bulletin* (new series), October 1934
- Even edges out to the columns
- Our first anniversary
- 1933- 1934 Council depleted in members

- The publicity value of the *Journal*
- An appeal against prejudice
- P Cleator and Grindell Matthews
- Pirquet and books in the German language.
- Askham on interplanetary radio communication
- The possible significance of long delayed radio echoes
- The "British Rocket Society"?
- The American Rocket Society
- J G Strong 40 years later
- Members vote for original name of the BIS
- Pirquet joins the Society's Information Bureau
- Mr James Free Jr Assistant Hon Secretary
- "Moon Landing" members' notepaper is designed
- An editorial committee is set up
- Mr James Free Jr Hon Treasurer
- The Council of the Society is brought up to full strength
- The *Bulletin* (new series) November 1934.
- The President leaves early.
- Norman Weedall, Hon Librarian of the Society
- Various suggestions made at a meeting
- Decisions made re the *Journal* and the *Bulletin*
- Plotting a path to the planets
- The question of the headquarters address of the Society
- Divergent points of view become evident
- Reasons for the onset of dissension
- The battle of the addresses
- Mr P Cleator - not very communicative
- The "Secretary" or the "Hon Secretary"?
- The *Journal*, a bone of contention
- Material and monetary contributions to the BIS by Cleator, Askham and Johnson
- New members, Arthur Clarke and Eric Russell
- A Clarke's telescope etc.
- A Clarke's lecture on "Life on Mars"
- A Clarke's article, "The Problem of Mars" fails to appear in the *New Columbus*

- A Clarke availed himself of services of the information bureau
- A Clarke and the speed of light
- A Clarke and the members' notepaper
- Walter Gillings supplies Clarke with science fiction magazines
- Eric Russell and interplanetary communication (the *Ida and Victoria* magazine)

THE YEAR 1935

Apart from the difficulty of the Society, having been governed by a Council depleted in numbers, the Constitution that had been drawn up did not make it sufficiently clear as to which decisions should be made on a vote of the general membership present at ordinary meetings of the Society, and which decisions should be taken only by the Council of the Society, gathered in solemn conclave. This position has led to some confusion, and was at least in part, the reason why the Society publication, the *New Columbus* was first produced - then never issued.

Physically, the *New Columbus* was half foolscap in size (6.5 x 8.25 inches) and mimeographed with dark blue ink on a pale blue paper. The single column transcript was even on both sides of the column, and the cover and inside illustrations were by Eric Russell who was joint editor with L Johnson. The cover illustration was in fact a rehash of a spaceship taken from the cover of an issue of the American science fiction magazine *Astounding Stories*. The science fiction influence was also evident above the editorial page, which boldly proclaimed that "Today Fantasy-Tomorrow Fact".

The contents were as follows:

Editorial (by Frank Russell)

Interesting Literature

Questions and Answers Department

Thanks W. S.!

Discourses by "Astrovox" - "Reason and Emotion"

"Sense and Nonsense"

Notes and News

Meetings of the Society

Correspondence with the Society

Coming Features

The Society Library

Particulars of the membership, etc.

Coming Features included:

Interplanetary communication by "Astrovox"

The Ethical Slant by "Astrovox"

The Elements of Astronomy by "Cosmonaut"

The Problem of Mars by Arthur Clarke

Cosmic Menaces by W Dunbar.

"Astrovox" and "Cosmonaut" were both pseudonyms of Eric Russell. I know of only two copies of the *New Columbus* in existence today, both copies being in my possession, the second one, having been left to me by my late friend, Norman Weedall. But there may possibly be further copies extant, as I sent a number to Edward Ted Carnell, in 1936. As the process of duplication was never completed, the last three pages of each copy are blank, and were to have contained notes and news and particulars of membership, etc.

Of some relevance are the editorial and Eric Russell's reports on ordinary meetings of the Society, which were held on 30 November, 14 December, 28 December 1934, respectively. These items are reproduced as follows:

Society publications

During the past year or so it has been the ambition of the officers of the Society to provide members with as much literary matters pertaining to interplanetary travel as they could, always, of course, commensurate with the dictates of economy.

With this idea in mind, the monthly report was issued, and this was recently enlarged to form the *Bulletin* of the Society. These publications, of course, were primarily intended as a supplement to the printed *Journal*. By December, however, we had reached the stage where it was necessary for us to make a choice between the duplicated *Bulletin* (issued monthly) and the printed *Journal* (issued quarterly). It was, indeed, very doubtful as to whether the *Journal* could be afforded in any case.

Last year, the bulk of the cost of publishing the *Journal* had been very kindly borne by Mr J Moores, and in the coming year we would most probably be left to our own resources. Besides the expense of the publication of the *Journal* the overall cost of managing the Society was extremely high considering the number of members we possessed.

Former *Bulletins* and notices had been produced on a duplicating machine in the office where one of the members was employed, but this could not be continued indefinitely. Thus, arose the proposal that the Society should purchase its own duplicator.

The Council, after very careful consideration of all the facts of the case, decided that the only course that could be pursued lay in the purchase of duplicating machine and discontinuation at least temporarily, of the printed *Journal*.

The new *Bulletin* would be produced by Mr L Johnson (Hon Secretary of the Society) with the aid of Mr E Russell (member-Liverpool) and the number of pages, duplicating etc was left to their discretion. Thus, it is that you perceive the first issue of a series of new *Bulletin*.

Members will receive copies of this publication as regularly as the inflow of subject matter permits. The reason for any doubt remaining as to the regularity of issue is due to the fact that we now have to contend with the reversal of the old state of affairs - instead of us not having enough space at our disposal, we now have almost as much as we could wish for!

We therefore appeal to all our members to do whatever lies within their power to supply us with suitable material for publication.

It was suggested that the *Bulletin* of the Society should be given a suitable and descriptive name, and the meeting of the Society on 28 December 1934, the title of the *New Columbus* was decided upon by popular vote. We would, however, like to learn the opinions of the rest of the members on the subject. We would appreciate replies to three questions: (1) Is the name desirable? (2) Can you suggest any more suitable appellations than the N*ew Columbus*? and (3) Can you supply us with better cover design?

Your guidance, the organ of the American Rocket Society is entitled Astronautics, and the publication of the Cleveland Rocket Society is called Space. Any other criticism of this publication will also be welcomed by the editors, as of course we are doing our utmost to provide our members with a type of publication they themselves, prefer. - Editor.

Meetings of the Society reported by E Russell

Meetings of the Society, which are held every other Friday, took place on the following dates in 1934 - 30 November, 14 December and 28 December.

Ordinary meetings of the Society

30 November

The chair was taken by the President, Mr P Cleator.

The minutes of the meeting held on 14 November were taken as read.

Forging the literary link

Mr Secretary Johnson reported to the Chair the result of his enquiries for a portable printing machine or duplicator. The Society is feeling the need for a method of producing its literature and publicity matter cheaply and impressively, without recourse to the outside agency. Members were very interested in the examples of work done by the various machines. It was decided that the matter be referred to the Council to be considered at their next meeting.

To Cure the Curious

As an addendum to his report Mr Secretary Johnson suggested that future bulletins be produced in excess of requirements, the surplus to be used to soothe those correspondents not yet initiated in our mysteries.

Notepaper of Note

From the meeting, via this page, the members expressed their thanks to Mr K Chapman for his excellent design for the Society's notepaper. Mr Chapman is to receive a small supply of his own handiwork as a token of appreciation.

Post Haste

Mild interest was created by the news that the French Air Ministry has authorised Herr Gerhard Zucker, German rocket experimenter, to shoot a rocket southward across the Channel. It is stated that 12,000 letters, will be transmitted by this process, and tests are to be started shortly. Herr Zucker carried out demonstrations during the International Air Post exhibition in London last May. He fires a rocket 17 feet long by three feet in circumference. This size of rocket travels, about five miles; one which larger would have to be employed to traverse the Channel.

A voice from Vienna

A lengthy and encouraging letter from Ing. Baron Guido von Pirquet was read to the assembled members by Mr Secretary Johnson. Baron von Pirquet dealt with the subject of a recent debate, "The Temperature in Space", and sent a well-drawn graph to illustrate his reasoning. With this letter, the famous mathematician unconsciously drew attention to a point we wish more of our distant members would keep in mind - it is a simple matter for such members to attend any of our meetings by proxy.

The Hon Secretary Goes Seal Hunting

It was stated that some members had asked for supplies of adhesive seals alike to those of the science fiction league. The Hon Secretary was requested to make further enquiries about the demand, design and price thereof.

Mail Mags

Members gave further consideration to the problem of circulating the library through the membership outside Merseyside area. They decided that all magazines and all books in the library would be sent to distant members upon application accompanied by stamps for postage. The Hon Librarian was requested to make a list of all the publications in his possession. This to be inserted in the *Bulletin* for the information of those concerned.

The "Windsor" wonders

A short but satisfying discussion on "The Possibility of Life on the Planets" followed by a reading by the Hon Secretary, of an article upon this subject which appears in the November issue of the *Windsor Magazine*. It was the general opinion that the author of this article failed to take into account the very great likelihood of our neighbouring planets evolving their own forms of life. Mr K Gibbs does not share the author's doubts for he sent the meeting an enthusiastic letter accompanied by a drawing of rocket-stabilisers.

Ordinary Meeting of the Society

14 December 1934

The chair was taken by the President, Mr P Cleator.

The minutes of the meeting held on 30 November were taken as read.

Welcome! Mr Stroud

Mr Holden Stroud of Belmont, USA, applied for associate membership and was elected.

Buttons put in Holes

Mr L Johnson said it was desirable that the members arrived at the decision upon the matter of lapel buttons.

In answer to Mr C Askham he stated that 15 members had asked for badges. Mr Askham thought the number was not sufficient to justify further action being taken in the matter at the present time. He proposed that the whole question be dismissed, until such time as it was evident that the demand was greater. He was second by Mr T Ashcroft.

Satan Finds Work

Mr L Johnson unconsciously asserting the rights of his diabolical majesty, was responsible for long discussion which resulted in a somewhat more uniform distribution of the Society's work. In the recent past, Mr Johnson has had to deal with all the secretarial work as well as ninety percent of that required to produce the *Bulletin*. The members considered that the amount of work required had become too much for a single individual. The Council had already decided that Mr E Russell be appointed joint editor of the *Bulletin* (or its successor) with Mr Johnson. Mr Russell had not been backward in criticising the Society's publications and was promptly hoisted by his own petard. Mr P Cleator, President and Founder of the Society kindly consented to help Mr Johnson with the secretarial work.

The official address of the Society is now 34 Oarside Drive, Wallasey, Cheshire, England. All letters pertaining to the *New Columbus*, or other of the Society's published matter, should be sent to the editorial address, viz: 46 Mill Lane, Liverpool 13, England. It was agreed that the printed *Journal* should be dropped until such a time as circumstances justified its reappearance. The Council decided to purchase a duplicator suitable to cope with the task of turning out the Society's magazine for an undetermined time. This machine has been acquired with the financial aid of Mr Askham, to whom therefore the Society is indebted for the appearance of the magazine you are reading.

Invoking Kindred Spirits

Mr E Russell suggested that we had a potential membership within the ranks of other organisations bound by interests similar to our own. He thought it might be of advantage to get into touch with such people with the idea of a mutual exchange of literary matter, lectures, etc. He mentioned the Liverpool

Astronomical Society, as an instance. Mr Cleator promised to write to the local officials of this society and signified his willingness to give them a lecture upon 'Interplanetary Travel 'if their members proved to be interested. Mr Russell also suggested that the details of our Society be sent to Messrs Kellys for insertion in the local directory. Mr Johnson said that our name did not appear in the red book and that he would take steps to notify both publications.

Let's Dump Debt's Lump

Under the heading of "any other business" Mr Russell raised a rather interesting point - he asked that the Society look into the question of buying the Secretary's typewriter.

A short time ago, owing to the amount of work in connection with the production of the monthly report or Bulletin, and attending to the ordinary business of the Society, it became necessary to buy a new machine - the old typewriter being entirely unsuitable for the whole work. This the Hon Secretary paid for out of his own pocket.

Mr Russell suggested that as the machine was being used entirely on the Society's behalf, it was rather unreasonable that the Hon Secretary should have to bear the full cost. This was objected to by the President and the Vice President, Mr Cleator and Mr Askham for the following reasons - that the Hon Secretary had not asked the opinion of the Council on the matter; that in their opinion agreement would set a dangerous precedent in that anybody who, at any time, had reason to do work for the Society could buy whatever was needed, use it for their own personal affairs, then come to the Society and demand payment.

Mr Russell thought that the Society was capable of deciding what expenditure was reasonable and what was unreasonable. He moved a resolution (seconded by Mr N Weedall) that the Council interview Mr Johnson with the object of reaching a satisfactory settlement. To this the Council agreed, and the matter was left thus. (Report by Norman Weedall).

The *New Columbus* Wants Cargo

Mr L Johnson stated that he was still receiving enquiries for the Society's publication from non-members and for extra supplies from members. He had gone into the question of costs and proposed that future issues of the magazine be made available at sixpence per copy, or $1.50 post-paid in the USA for all non-member subscriptions. This was approved. Mr E Russell remarked that he feared the joint editors would have to face a shortage of suitable material in the future if the magazine was to appear every month.

He knew that a few technical articles were in possession of the Society and thought that, as they were written by fellow members, they should be placed at the disposal of the joint editors. Mr Johnson said that in the next issue of the magazine, he would invite members to submit their contributions to the Society's organ.

Getting our own back

An article in the Armchair Science entitled "Interplanetary Radio Echoes and Their Use in the Exploration of Interplanetary Space", was read to the assembled members by Mr Johnson. Briefly, it described how the motions of the heavenly bodies, their distances, etc. could be told by means of the time lag between the emanation of the radio signals and the arrival of their echoes. Time did not permit much discussion upon this subject as it appeared as the last item on the agenda.

It had been intended to hold the next meeting upon 28 December, but the Council members were doubtful of their ability to attend on that date. The members were in favour of a meeting be held so, it was decided to hold it as an informal meeting if the Council could not manage to get there.

Ordinary Meeting of the Society.

28 December 1934

The chair was taken by Miss Hastie.

The minutes of the last meeting were taken as read.

Who's to Rah-rah Prodigals?

Mr J Davies raised the question of whether election of new members should be considered at these meetings or whether it was the prerogative of the Council. Mr L Johnson said that it was difficult to get a complete attendance at Council meetings owing to pressure of personal affairs upon the various members. Mr Davies replied that he appreciated the difficulty but wished to question the correctness of the procedure. Mr E Russell suggested that the procedure was quite correct in view of the fact that the Council had left it unchallenged up to that time. If the Council objected it was up to them to take action and, until they did, their silence could be construed as consent.

Wits in a name

The members gave further consideration to the various suggestions for improvements in the *Bulletin*. Mr Russell proposed that the *Bulletin* be given a title. Mr T Ashcroft seconded, and the proposal was carried by a vote of those present. Mr J Davis said it was not clear about why a title was considered necessary. Mr Russell replied that it was a duty of Mr Johnson himself, as joint editors of the next issue, to make it as attractive and imposing as possible within the limitations of the time allowed and the finances that are available.

He thought that a title would lend to the publication's much needed personality, that would add to its dignity - especially if accompanied by a suitable cover design. Mr Russell produced a titled cover design which he had drawn and submitted it for approval of the members. After examination, the members voted in favour of his design - which now appears upon the cover of this issue.

Mr L Johnson, read out a list of various titles thought to be suitable for the post for the purpose in mind. The members voted in favour of The *New Columbus* this being the title used by Mr Russell in his cover design. *Cosmonautics* ran second, the members considering it a good title, but which had too strong a resemblance to the American Rocket Society's *Astronautics*, as well as to M R Esnault-Pelterie's recent book of like name. Mr Russell remarked that in deciding matters of this description, the editors wished to be ruled by a majority vote in the general membership.

There was not sufficient time to ascertain the wishes of distant members before the next issue was due to appear. It was unfortunate, but they thought that the vote of the members present will be fairly representative of the whole. Mr Johnson would write an editorial in which he would appeal to distant members to send us their criticism of the new magazine the policy of succeeding issues will be decided by the response to the editorial.

Cleveland Whets Our Ambitions

To illustrate the standard of publication, which the joint editors were anxious to equal, and if possible beat, Mr Russell produced copies number four and five of *Space*, organ of the Cleveland Rocket Society. He said that these two magazines, together with four photos, had been sent to him in response to requests for material to insert in his serial, which is now running in the *Ida and Victoria* magazine. Members were very interested in the photographs which showed views of the Cleveland Rocket Society's rocket field, the proving stand (apparently constructed of steel tubes), a rocket motor in action, and the shielded observation trench.

Number four of *Space* was particularly well illustrated with drawings, photographs, and even humorous sketches and both copies contained a highly satisfactory amount of technical literature. Mr Russell said that the Cleveland Society were preparing for another series of experiments with a newly designed motor. It was likely that the members would have details of these experiments towards the end of January. He said that when he had finished with these two publications he would put them into the files of the Society's library, so that they would be available to the general membership.

A Round-up in Texas

Three very interesting letters from Mr B Goree of Texas were read to the members by Mr Johnson. Mr Goree imparted the news that he has a major share in the first roundup of general material dealing with rocketry in all its aspects. News items, details of experiments and technical articles were to be incorporated in the book to be called *Handbook of Rocketry*. Mr Goree suggested that the Society take a number of the handbooks, either by purchase or in exchange for an equal number of our publication. Mr Johnson, painfully unaware of whether the *New Columbus* will entice brickbats or bouquets, said he thought it would be best to leave Mr Goree's offer in abeyance until the January number of our magazine has appeared and Mr Goree is able to form an opinion of it.

Will you Preach Sedition?

During a discussion about methods of increasing the Society's membership, it was found that some members of our Society are also members of other societies, with kindred interests, or are in touch with organisations of this type. It was thought desirable that, in our report, an appeal be made to these members to make use of their connections on our behalf. There are recruits to be drawn from the ranks of various astronomical, scientific, all engineering and radio organisations, and it will be of great help to us if those of our members who belong to such bodies can bring us some of their fellow members - after interesting them in spatial science.

The head of Science - and the Bailiffs!

Mr L Johnson reported that the Society's debts were, at the moment, slightly in excess of the cash in hand. Several subscriptions were due, and payment thereof would be welcome. The response of the members present was sufficient to meet bills to be paid in the coming month.

A Terro-Martian War?

An article entitled "Could a Manned Rocket Reach Mars?" by Henry Norris Russell, was read from recent issues of Scientific American. Throughout the greater part of his contribution the author dealt with outsize bullets, or projectiles, and confined the subject of manned rockets to the small, final paragraph. Members' criticisms of the articles were numerous the general opinion seemed to be that the author knew a lot about planetary motion, but not much about rocketry.

Mr L Johnson announced that the Annual General Meeting of the Society would be held on Friday, 11 January 1935. Miss Hastie whose chairmanship had enlivened the meeting declared the assembly ended.

With Arthur Clarke complaining bitterly to me in our voluminous correspondence that he had received no Society publications for over three months the question as to whether the *New Columbus* would ever see the daylight dragged into the month of February 1935 (although the first issue of the publication was dated January 1935). Russell was still working on a cover illustration for number two and I was preparing to submit Russell's *Ida and Victoria* magazine series *Interplanetary Communication* to the Council for possible inclusion in a future issue.

Having flown to Berlin in July 1934, and seen the German rocket experimenters, Cleator was on a second visit in January 1935 while the future of the *New Columbus* was in the melting pot. No doubt his return to England and his resumption of his normal interest in Society's business resulted in a decision finally being taken by the Council, with regards to the publication.

The upshot was that before the end of February all members were circularised expressing regret that no Society publication of any kind had been issued since November 1934, explaining that the Council had the whole matter on Society publications under review and stating that members would be informed of the Council's decision in due course.

What the notifications to the members meant (but what they were never told!) was that the Council had finally decided to "ban" the *New Columbus*, and that the publication would be sunk without trace in spite of the fact that it was almost ready for distribution.

There was no doubt (although I say this myself) but that for my mimeographed publication, the *New Columbus* had been excellently produced, but the Council's decision had been based on their view that the whole tone of the publication was far too flippant to represent a Society such as the BIS had pretensions to become. The flippancy was mainly evident in Eric Russell's reports on the meetings of the Society, and in the small cartoon reproduced from *Pearson's Weekly* showing a small boy aiming a rocket at an apple tree, and the cartoon being entitled "The Public Enemy".

In retrospect, I can understand the decision of the Council not to issue the *New Columbus*, although it was a pity that so much hard work by Eric and myself had gone into the abortive publication. Eric had tried to make the report on meetings popular in tone and had succeeded only too well. It was also perhaps unfortunate that some decisions were taken at the informal ordinary meeting of 28 December 1934, that might better have been taken by the Council itself.

My feeling now is that little harm could have been done to the cause of the Society and of interplanetary travel if the publication had been issued to members only rather than leave our efforts to have been wasted altogether.

Arising from the *New Columbus* incident, therefore, no issue of the *Journal* appeared dated January 1935. When an issue did finally appear dated May 1935 (once again edited and at least partially paid for by P Cleator, it reverted to the format of volume one, number one (January 1934) consisting of only six folding octavo pages. The content comprises an editorial by P Cleator on "Extra Terrestrial Life" and an "Analecta", with details of membership on page six.

The front cover showed a photographic view of the planet Jupiter (the block having been lent by Eric Russell), and Chambers' Journal had lost its free full-page advertisement. By this time, no publication of any kind had been issued by the Society for over six months.

A long letter from Russell early in February went into the legal aspects of the Society's constitution. Said Russell: "Cleator's determined dismissal of members' proposals has no legal standing. If the Annual General Meeting was a proper meeting, despite his absence, he is obliged to stand by the decisions made at that meeting. If, as he claims, it was not a proper meeting, then the proper meeting has yet to be held, and he has no right to determine how future meetings should be conducted until that meeting is held a proper constitution has been properly drawn up and the Council have been elected and invested with the powers they have already taken on to themselves. The members, I am afraid can do nothing in view of the Council's dictatorial action, so long as the Society has no legal status. In brief, the Society's affairs are in a helluva mess...."

Steps were taken to rectify this position, and by the middle of July I was able to submit a new draft constitution to Cleator together with an invitation to him to suggest any amendments he might consider necessary, so that the matter could be submitted to the Council three weeks later. But the problem of arranging for the Council to meet appeared to be as difficult as the formulation of the Constitution itself.

In the meantime, at the instigation of a schoolmaster, Mr Francis HP Knight (member - Walsall) the Society took a paternal interest in IDO (reformed Esperanto) as an international language, and an article that I wrote called *Space Rockets* (or *Space Fuzeili!*) concerning the aspirations of the BIS was published in the *Ido Magazine*. I went on to exchange 100 sheets of member notepaper for a number of Mr Knight's surplus science fiction magazines.

Referring to the dissension of a Society, whatever factions may have existed (and one might say there was a dictatorial faction and a democratic faction), the divisions were not always clear cut and it is true to say that all had the good of a Society at heart in their various ways. No real personal animosity existed at the time and any one faction did not necessarily always oppose the other faction.

There were, what today might be called "floating voters", and all parties frequently got together in agreement. For instance, Eric Russell (despite whatever his feelings may have been about the ban on the *New Columbus*) was it quite willing to lend Cleator printers' blocks of his own that Cleator could use when he resumed as editor of the *Journal*, which was with the May 1935 issue.

Russell later also arranged for similar blocks to be made available for use in Cleator's book *Rockets Through Space*, although in the event Cleator, did not need to use them. The May 1935 issue of the *Journal* was unfortunately put out in a considerable hurry because of the long silence that had ensued since November 1934.

Early in February 1935, the interest of Society members had been enlivened by the advent of Herr Willy Ley, en route from Germany to the United States of America, a fugitive from Hitler's Third Reich. Apparently, Cleator's visit to Dusseldorf in January 1935 had been part of an attempt to assist Ley out of Germany, but Cleator had had to return to England without him.

Whatever dissension may have existed within the British Interplanetary Society it never reached the stage that it had in the German Interplanetary Society (Verein fur Rhaumschiffahrt) or in the Manchester Interplanetary Society, both of which, for various reasons had split respectively into two separate organisations. Willie Ley's group had left the VfR to join the already existing E.V. Fortschrittliche Verkshrtechnik (Society for the Progress of Traffic Technique) and the Manchester breakaway group to form the Manchester Astronautical Association.

However, this split in Manchester was not occasioned by anything or anyone as formidable as Herr Hitler. So it was at the instigation of P Cleator, Willy Ley, who had been one of the leading lights in the German rocket scene was invited to the United States of America by G. Edward Pendray President of the American Rocket Society and by Edward Hanna, (Founder of the Cleveland Rocket Society) both of whom were eager that the USA should become the first country to establish a regular rocket mail service.

The Liverpool Echo dated Friday, 8 February 1935 carried banner headlines:

> When Rocket Carrier Succeeds
>
> US to England in 30 minutes
>
> Rocket Pioneer in Liverpool
>
> Jules Verne Touch
>
> Science Triumphs of the Future.

The Echo added that Ley was staying as a guest of Mr P Cleator whilst he was in England and was to address the British Interplanetary Society that evening.

Alas for the word "address"! Willie Ley certainly attended the meeting of the BIS that evening, held as usual on the first floor of the Hamilton Cafe, 56 Whitechapel Liverpool - not a very pretentious rendezvous, but very convenient and suitable for the needs of the Society, such as they were at the time.

Ley certainly gave a very interesting lecture to the Society. Three reporters, and a photographer were present - plus a dozen members of the BIS! Oddly, it was the reporter from the Liverpool Echo, very unimpressed by the whole proceedings and with the relatively few members present, who left in high dudgeon. The remaining reporters stood their ground manfully, and reports and photographs duly appeared during the course of the next few days in the Daily Graphic, the Daily Sketch and the Sunday Graphic. All three of these papers, unfortunately, are now defunct.

In the photographs, Willie Ley is shown, surrounded by BIS members; he is clutching his "plans" for sending rockets to the moon and explaining exactly how this feat is likely to be accomplished. Unfortunately, the plans he was holding in his hand consisted only of a blank sheet of foolscap paper!

The BIS members shown in the press photographs can be identified as P Cleator, Miss Eileen Hastie, J Free Jr., J Davies, C Askham and myself, as Miss Norah Bellew, a non-member lady friend of mine who had accompanied me to the meeting.

As mentioned earlier in this narrative, No 56 Whitechapel still survives as "Tape" Electronics on the ground floor and as Sukie's Hair Stylist on the first floor where we actually used to meet. The premises (as well as ourselves) however, nearly met an untimely end at about the period of Ley's visit, when we were so engrossed in our deliberations one evening that in spite of the noise and commotion prevailing downstairs, we did not realise that the premises were on fire.

While comments were made by the members present with regards to the likelihood of a drunken brawl persisting downstairs in the remainder of the cafe, or perhaps in the street outside, we did not immediately realise the reality or indeed the desperate nature of our situation. Not until the door burst open and smoke-begrimed fireman burst in brandishing an axe, did we come to an understanding of what had been happening. And the fireman himself was as astonished as we were to find us oblivious of the fact that downstairs portion of the cafe had been gutted whilst we were intensely debating how rocket ship might reach the Moon!

As luck would have it, the wide staircase, led (and still leads) straight from the street, up to the first floor, so, as the, the fates would have it we no doubt could have made our exits without having to pass through the lower section of the cafe at all.

Following the demise of the *New Columbus*, G Edward Pendray made a proposal that the publications of the three chief English speaking societies, The American Rocket Society, the Cleveland Rocket Society and the British Interplanetary Society should be merged into one publication. In this way, it was hoped that a well-illustrated monthly magazine could be produced at a minimum expense to each organisation.

This possibility was explained at length by P Cleator in his editorial to the May 1935 issue of the *Journal*, who went on to say that each of the three societies would need to contribute a yearly sum of not less than £50, and that we, at least, could not afford anything like that amount.

In the meantime, Cleator had to fall back on the old dictum that whilst it was hoped to issue a truncated *Journal* at more frequent intervals, still, this would only be often as funds permitted. He went on to urge upon existing numbers, the vital importance of introducing prospective members to the Society wherever and whenever possible.

The state of the Society, and the state of the interplanetary "scene" - such as it was - was mirrored in the correspondence, in which I indulged during this period and later, notably with Eric Russell, Arthur Clarke, Francis HP Knight, as well, as of course, with P Cleator and (later in the year), Walter Gillings. In fact, Arthur Clarke went so far as to describe himself (truly) as my most persistent correspondent.

This was only the start of a phenomenon that was to snowball in the years before September 1939 when the outbreak of the Second World War put us all into hibernation. It was in an undated postcard that Eric Russell wrote to me asking for Arthur Clarke's address. Russell wanted permission from Clarke to print his article, "The Problem of Mars" in the *Ida and Victoria Magazine*. In his letter to me, dated 11 March 1935, Arthur Clarke stated that he had heard from a member called Eric Frank Russell, who wanted permission to print his article on Mars in a private magazine. He would have preferred to have had his article published in the *Journal* of the Society, and in fact, he then sent another article, as he presumed we intended to publish further issues of the *Journal*.

I explained to Arthur that I had read the article to members at a recent meeting, and that Russell would be using it in the *Ida and Victoria Magazine* in the near future. As some kind of consolation, I went on to assure him that we would soon be forming an "experimental committee". When he finally received the six-page May 1935 issue of the *Journal* Arthur wrote complaining that it seemed a little small. He thought that perhaps a smaller type would enable better use to be made of the same amount of paper, but I hastened to inform him that it was basically the amount of typescript that determined the cost of publication. He hoped that reports on meetings and copies of lectures given by members would continue to be available. He also expressed an interest in the library of the Society.

He was a bit out of luck because in the future none of these facilities was to be made available to members residing outside the Merseyside area. Brief reports of the meetings would continue to appear each month in *Practical Mechanics,* and it was hoped that the next issue of the *Journal* would consist of eight pages. This hope was reinforced by the fact, as I mentioned to Cleator, that we had a balance of £2-10-0 (£2.50) with all bills paid.

Colin Askham thought that a new printed edition of the *Journal* would be put out as soon as possible, and hopes were raised further by the news that the Ripley Printing Society - recommended to Cleator by Russell - had submitted an estimate showing a saving of one whole pound sterling on our previous costs.

And a ploy was also made to obtain printers' blocks free of charge in order to illustrate the *Journal* by using those that Russell had incorporated in his series of articles in the *Ida and Victoria Magazine*.

Thinking that the existing design looked a bit old fashioned, by July 1935 I had designed a new style of members notepaper based upon that of the Rationalist Press Association and flaunting the slogan

"Devoted to the Conquest of Space". This slogan was not much to the liking of Phil Cleator, who thought it sported religious connotations, despite my own view that one could be devoted to objectives other than religious ones.

The new members' notepaper was certainly very comprehensive in design but while Cleator had definitely used the old-style notepaper, strangely, I have no record of him ever having used the new product. The quarto-sized sheet included a column down the left-hand side, giving the names of the officers of the Society, as well as the names of the Founder Fellows, and the Honorary Fellows of the Society.

Founder fellows were C Askham (G6TT), H Binns Jr., P Cleator, J Davies (G20A), W Dunbar, J Free Jr., L Johnson, T McNab and N Weedall. Honorary Fellows were R Chambers, Fraulein I Kuhnel, Willy Ley Dipl., Ing.E. Loebell, Professor A Low, G Edward Pendray, Dr Jakow Perlmann, Ing Baron Guido von Pirquet, N Moore Raymond, Ing. F Schmiedl, Ralph Stranger, Raymond Thiele, Richard Thiele and Mrs. A Weston M P S, the name of Ralph Stranger being added to the list in early 1936.

In consultation with P Cleator it had been decided not to include the names of the contemporaneous organisations, which would have included such as the American Rocket Society, and Cleveland Rocket Society.

By August 1935, it was clear that circumstances had put P Cleator back in the driving seat, but Askham's interest in the Society appeared to have waned considerably. In the meantime, Cleator was working on his book on interplanetary travel, at that stage to be entitled, *The Rocket Era*.

In the same month, a sensation was caused by the announcement of the discovery by Professor E Appleton of a layer in the ionosphere (dubbed "Appleton's Inferno" by P Cleator), allegedly with temperatures ranging up to 1000 degrees centigrade, and which - to the delight of the opponents of the interplanetary idea was "bound" to prohibit any idea of space travel, or indeed to any great height in the Earth's atmosphere.

Arthur Clarke was furious over an article on this subject that had appeared in the Daily Mail, and he insisted that the arguments against the possibility of space travel did not hold water (if that is the right expression to use!) for at least two reasons:

(1) 1000 degrees is not so hot after all, and asbestos, etc could protect prospective space travellers, until they got into free space;

(2) The gas in the ionosphere is so rarefied that even if it was 1,000,000 degrees it would have comparatively little effect.

Clarke went on to say that he believed that Sir James Jeans mentioned that the galactic nebulae such as the one in Orion are at very high temperature, but as one would have to travel 100 yards or so in them before you hit a molecule it didn't matter very much!

He went on to urge that either Cleator or myself should write to the Daily Mail, or if humanly possible get a proper article accepted by the newspaper, refuting these arguments. He went on to say that Appleton was a brilliant man, and he suspected that most of the deductions given were embroidery by the Daily Mail.

I wrote back to Arthur stating that I had traced the original article by Professor Appleton in *World Radio*; I added that it was obvious that the Mail had jumped to hasty conclusions, and that I proposed to write to the editor in order to refute what they had declared concerning the proposed lunar voyage.

The only upshot of my letter sent to the Daily Mail was that I received a formal acknowledgement - which was perhaps more than I had anticipated receiving in any event.

Considering that it was the height of the summer, the month of August 1935, saw an unaccountable amount of activity of interplanetary interest, and this apparently stimulated Walter Gillings to revive his interest in such activities. He wrote to enquire how our liquid fuel rocket experiments were progressing, and whether he could acquire information in the form of an interview with our president.

He realised that Mr P Cleator "did a bit of journalism" (as he put it) and didn't wish to tread on his corns, but if he could get a really good story on the Society, he might be able to get into the London papers. Gillings, incidentally, was at the time, a reporter on the Ilford Recorder. In reply, I went on to supply Gillings with a tremendous amount of information, not only about the Society, but also my personal history, adding that having started a Society with only ten members we then had about 80. I added that I was sure that Phil Cleator would be only too willing to help him in any way that he could.

The idea of Gillings enquiring about the progress of the Society's liquid fuel experiments was one of the best jokes I had heard for some time, for we were as far away from conducting experiments as we were in miles from the Moon!

And I am afraid that I very much misjudged Cleator's attitude towards Gillings' request, for Cleator hit the proposition on the head, stating that he was not anxious to obtain what he called 'cheap publicity.

Feeling that Gillings was sufficiently responsible to be able to arrange suitable publicity, I sent him the draft of an article I wrote out for him, headed "The Flame Belt Will Not Impede Interplanetary Travel" – with proposed subheadings such as "Interplanetary Society Unperturbed" and "Future Spaceports in the Arctic?."

Gillings thought my letter to the Mail was far too technical for the Mail – and also for Walter Gillings! He asked for a detailed report on Appleton's Inferno, which I did for him on a question and answer basis. I added that Ralph Stranger, in an article in World Radio, commented upon the scare headlines used by journalists about the temperature of the ionosphere and about the so-called obstacle to interplanetary travel – which he seemed to think was non-existent.

Unfortunately, the onset of the elections took precedence with Gillings over the problems posed by the temperature of the ionosphere, and as far as I know, the information I had supplied to him was never put to any particular use.

In spite of the furore created by the discovery of Appleton's Inferno, the BIS went ahead preparing an eight-page issue of the Journal (edited by P Cleator) which finally appeared dated October 1935. The eight pages, strangely enough, opened out into one very large sheet, which when folded appropriately, yielded an eight-page Journal of a rather large Octavo size.

On the front cover appeared a photograph of Experimental Rocket No 1 of The American Rocket Society. Contents: Editorial by P Cleator; "Why not shoot Rockets?" by Edward Pendray; the inevitable Analecta, and on the back cover a preview of Cleator's forthcoming book, now, re-entitled *Rockets Through Space*. This showed the scene from the Ufa film, *The Girl in the Moon* with a spaceship ready for launching.

About this time, I received a letter from Eric Russell, inviting me to send a copy of the current issue of the *Journal*, and any back numbers to one Dr W Olaf Stapledon, a professor at Liverpool University and an author in 1930 of *Last and First Men*, a science fiction epic. Unfortunately - or otherwise, as the case maybe he had never heard of science fiction apart from HG Wells, until Eric Russell and myself, visited him and (were royally entertained) at his bungalow at Caldy Hill, Wirral, Cheshire (now Merseyside Metropolitan County) and where Stapleton's widow resides to this day.

Among Stapledon's masterpieces were "Last Men in London", "Odd John", "Sirius", "Old Man in a New World", "Death into Life", "The Flames", "Worlds of Wonder", and "A Man Divided" - not to omit "Star Maker".

There is no doubt that, that if anyone honoured, the BIS by becoming a member of the Society, it was Dr William Olaf Stapledon. And I see in "Who's Who in Science Fiction" by Brian Ash that one of Eric Russell's greatest achievements, was to unveil the world of science fiction and the BIS to Olaf Stapledon.

The efforts of Russell and myself, must have borne fruit, because a letter dated 7 August 1935 to me from P Cleator stated that a new member had come in - "a Mr Olaf Stapledon of West Kirby". Cleator added that he had had no record of having had any previous communication with Stapledon; would I send a receipt for half a guinea (10s 6d or 52½p) representing the amount of annual subscription for ordinary membership. Cleator, who apparently had no idea who the new member was, added that he proposed (quite correctly, of course) to retain the money to cover the postal expenses.

In spite of the furore occasioned by the discovery of "The Flame Belt", in spite of the publication of the eight-page October 1935 issue of the *Journal*, and in spite of the enrolment of Dr Olaf Stapledon, the year 1935 seem to be ending on an indeterminate note. The trough of despond into which the Society seemed to have sunk was fortunately enlivened by the prospect that a Liverpool chapter of the Science Fiction League(sponsored by the magazine, *Wonder Stories*) would shortly be formed, and that the chapter members would meet on the same premises, and immediately following the BIS meetings. Such an arrangement was, of course, typical of the intimate relationship that existed in those early years between pioneers of the idea of interplanetary travel and the pioneers of science fiction.

In early December 1935. I wrote to Francis HP Knight (member - Walsall) to say that there was apparently little left in the BIS to inspire one to any effort. However, talks had been going on behind the scenes between some of the more prominent members, and it was apparent that the time was ripe for some sort of concerted action in order to revive the Society.

In the new year it was proposed to call a meeting, which should result in at least the publication of a regular monthly magazine on behalf of the Society, as well as the livening up of our other activities. Cleator's work on his book was most probably one of the causes of the Society's inactivity. It was hoped that all this would be changed in January, and particularly when the publication of the book would bring a much-needed revival of interest in the work of the Society. In the meantime, it was surprising that few members outside Liverpool had complained. Indeed, it looked as though followers of the cause of interplanetary travel possessed a remarkable degree of patience.

I went on to explain to Francis Knight that the situation in which we found ourselves was an exceedingly difficult one, and that it was extremely doubtful as to what would be the correct, and most advantageous method of dealing with our problem.

I possess copies of three reports that I compiled in March 1936 - the first, entitled

"Britain's Astronautical Pioneers" was sent to Professor A M Low, with a view to publication in the magazine *Armchair Science*, of which the professor was the editor. The second is "The British Interplanetary Society During 1935" and the third is "The Annual General Meeting 1936". One comment made in the reports was that the extent of the growth of interest in the possibility of interplanetary travel was well illustrated by the fact that Liverpool public reference library had - of its own accord! - recently included the word "Astronautics" as a classification in its catalogue of books.

In the reports, the problems relating to Society publications during the year 1935 were elaborated and it was clear that due in great part to the infrequent appearance of the *Journal* 1935 had been a poor year from the standpoint of an increase in membership. It had been gratifying, however, to note that the increase in the number of members, as such, at least showed signs of equalling the increase in the number of new associate members (those aged under 21 years of age).

When it is added to this that a disappointingly large number of members fail to renew their subscriptions, it must be imagined that the year had been a particularly unhappy one from the point of view of the Society. The only consolation was that the very reason for the paucity of support - the infrequent appearance of *the Journal* - had itself been instrumental in ensuring the production of a favourable financial statement. Although the financial position of the Society was nothing to boast about, nevertheless, it was some sort of achievement for us to have been able to show a surplus on the year's activities.

The formation during the year 1935 of an experimental committee and the establishment of a Research Fund had been accomplished more in an enthusiastic spirit of hopefulness rather than with any great faith in the possibilities of achievement. Meanwhile, in America, numerous experiments have been conducted by the rocket societies as well as private experimenters; similar work had been undertaken in France, Holland, India, Japan, and the USSR.

The reports included a description of the invention and construction of a spacesuit by Dr JBS Haldane and Sir Robert Davies for using stratospheric flights. This was in contrast to the announcement of the discovery of the so-called "Flame Belt" in the ionosphere. It was claimed that the spacesuit could carry a supply of oxygen sufficient to provide for a man on a journey to the Moon - if a vessel could be built capable of traversing the interplanetary void.

The interplanetary idea, I suggested in the reports, was rapidly gaining ground and was at last being recognised as a serious form of research and which would at least have a future in the construction of an efficient and speedy carrier of mail! At any rate, the attitude towards the science of rocketry by the vast majority of the general public had improved to that extent.

The reports went on to thank all those members of the Society and others who had worked so hard during the year 1935 to spread the interplanetary idea. The progress being made and that which was bound to come would go far to reward those enthusiasts for their very valiant efforts on our behalf.

Were it only to obtain the necessary support, the BIS could easily make its presence felt in the world of science, for astronautics had a future, such as was possessed by no other branch of research. The achievement of interplanetary travel (I went on) would be epoch making and would go down in history as the fulfilment of an age-old dream of the human race.

Once again, the American science fiction magazines had lent their support to the Society; letters had been published in *Wonder Stories* in January 1935 and in both *Wonder Stories* and *Amazing Stories* in April 1935. The publicity so obtained had helped enormously to offset the fact that only two issues of the *Journal* had been published, during the course of the year - and no issue of the *Bulletin* had appeared at all.

The report on the year 1935 concluded that "for many reasons, 1936 promised infinitely more progress than had its predecessor". And how right this prediction proved to be, for the year 1936 turned out in fact to be the most significant year of the Society's existence whilst it was based in Liverpool.

At the close of the year, a letter reached me dated 30 December 1935 from one "Edward John Carnell", Science Fiction League member number 1197. I, myself, could boast of having been SFL member number 383. Mr Carnell was writing to me, rather than Mr Cleator, because he was more familiar with my name, through my frequent letters that appeared in *Wonder Stories*. Carnell went on to explain that he was the London representative of George Gordon Clark's Brooklyn Reporter (a science fiction fan magazine) and it would be quicker for Carnell to communicate with me on behalf of George Gordon Clark than for me to have to endure the delays involved in transatlantic mail. Carnell was endeavouring to find out the extent of science fiction enthusiasm in this country, which he suggested was "extremely low", together with the news of rocketry and Esperanto.

Carnell went on to ask if I could supply him with any general news, news of experiments, or any short articles. In short, could he help me, or could I help him? He went on to say that he was 24 years of age, was not very well up on the facts of interplanetary travel, although astronomy and science in general, were of interest to him.

Little did I realise how my association with Ted Carnell was to develop over the years. Unfortunately, Edward John Carnell passed away in March 1972, but my correspondence with him (most of which took place before 1940) fills three folders and a total of about six inches in depth. Ted was also to play a leading part in the formation of the London branch of Society and the subsequent transfer of the Society headquarters to London, as will be explained in detail later in this narrative.

Summary of Events

- Decision making - difficulties imposed by the Constitution
- The *New Columbus*.
- I possess only two copies of the *New Columbus*
- Clarke complains re the paucity of BIS publications
- Working on volume 1 number 2 of the *New Columbus*
- Cleator returns from his third visit to Germany
- The Council regrets that no publications have been issued by the Society for three months
- The *New Columbus* is banned by the Council
- The *New Columbus* - "too flippant"
- Public enemy number No.1 - according to Pearson's weekly
- Russell's "too popular" report on meetings
- The *New Columbus* for members only?
- The *Journal* reappears dated May 1935
- Aspect of the new issue of the *Journal*
- The year, 1935 - a bad year
- More difficulties over the application of the constitution.
- A new constitution is drawn up
- Mr FHP Knight and *Ido* - Members' notepaper exchanged for science fiction magazines.
- No real animosity between the factions within the Society
- Russell lends printers blocks to Cleator
- Herr Willy Ley visits Liverpool
- Dissension in Germany and Manchester
- Herr Willy Ley invited to the USA
- The Liverpool Echo and Herr Willy Ley
- Herr Willy Ley addresses the BIS
- The Echo reporter leaves the meeting in high dudgeon
- Reports on and photographs of the meeting.
- The Hamilton cafe goes on fire.
- Publications of the English-speaking societies to be combined?
- The cost of producing a joint Journal £50
- The *Journal* to go back to square one - when funds permit

- Correspondence with Russell, Clarke, Knight, Gillings and Cleator
- Eric Frank Russell. What's the address of Arthur Clarke?
- Clarke hears from Russell
- Clarke found the *Journal* was "rather small".
- Hopes raised of an eight-page issue of the *Journal* £2 10-0 in hand
- An estimate from the Ripley Printing Society for producing the *Journal*
- Cheap blocks available through the *Ida and Victoria* magazine
- Appleton's inferno
- Arthur Clarke and Appleton's inferno
- A letter sent to the Daily Mail
- Walter H Gillings requested request details of Society's experiments!
- Gillings seeks to write a story of the BIS for London newspapers
- Cleator unable to help Gillings - dislike of cheap publicity
- The Flame Belt on a question and answer basis
- Information given is never used by Gillings
- The October 1935 issue of the *Journal* is published
- Dr W Olaf Stapledon
- Dr W Olaf Stapledon's works of science fiction
- Who's Who in science fiction
- Dr Stapledon joins the BIS
- Dull end to the year 1935
- Liverpool Chapter of the Science Fiction League is formed
- Hopes for 1936 in spite of the gloom of 1935
- Further difficult problems remain to be solved
- L Johnson to report on the year 1935
- Summary of the year 1935
- *Astronautics* in the Liverpool public reference library
- Favourable financial position at the end of 1935
- Experiments in 1935 all over the world
- The Haldane/Davies spacesuit
- Rockets for the delivery of mail.

- Thanks to all who helped in the interplanetary idea during 1935
- Hopes for the future
- Letters in the American science fiction magazines re the Society during 1935
- Greater hopes for 1936
- Letter from one Edward John Carnell
- Carnell acts as an agent for George Gordon Clark
- Carnell asks for news

THE YEAR 1936

Early in 1936 with the publication of Phil Cleator's book *Rockets Through Space* in prospect and the likelihood of a flood of enquiries with regard to membership of the Society, it was decided to print special membership application forms and details of the membership. One thousand of each of these items were to be obtained at a total cost of £2-7-6 (£2-37½p) and Cleator suggested that the only suitable illustration we had available for the details of membership form was the block that had been used for the first (January 1934) issue of the *Journal*.

When the account came in, the situation in which we found ourselves, was that we were £1 short of being able to pay! Otherwise, the "Research Fund" included a total of £3-6-9, of which 3s.3d had been subscribed by FHP Knight, and 3s.6d by member PS Hetherington. The balance of £3 was the amount we had owed Colin Askham who had paid the deposit for the purchase of the duplicator.

By the first week in March, however, funds had started rolling in again. The list of new members included one Mr J. Happian Edwards of South Chingford, who in his application for membership stated, "astronautics has been my life interest".

Mr J Happian Edwards certainly proved to be the most enthusiastic member of the Society. His thirst for scientific knowledge was probably only partially satisfied by *The Principles of Rocket Propulsion* by Herbert Chatley in *The Journal of the Royal Aeronautical Society* to which Cleator had referred to him in response to his request for further information concerning rocketry.

Edwards, then surprised us somewhat by writing to apologise for being unable to attend the Annual General Meeting in Liverpool (at which we did not expect to see him in any event!), but he added that he would like to be able to participate in any activities that may be taking place in London, or to be put in touch with any members in his vicinity who had knowledge of spherical trigonometry or lesser known facts of inorganic and thermal chemistry.

I had referred Edwards to Ralph Stranger's Science Review for information on rocketry and BIS activities in general and explained that as yet there were as yet no organised meetings in the London area. In fact, the extent of the membership in that area, did not yet satisfy the formation of a separate branch. However, I gave him the address of Frederick Addey, B.Sc. and J G Strong B.Sc. and I expressed the hope that something might possibly develop to our mutual advantage if we were able to contact these two members in the first instance.

Earlier in the year, Cleator had produced an excellent 12-page issue of the *Journal* (undoubtedly the best issue published up to that time), and furthermore had announced in his editorial that arrangements have been made with German Society Fortschrittliche Verkehrstechnik, whereby that Society would receive a supply of each issue of the *Journal* in exchange for a like number of their own publication Das Neue Fahrzeug". Copies of both publications would then be received from time to time by BIS members at no additional cost to them, and at only slight additional cost to the Society in printing extra copies of the *Journal*. A similar arrangement was in prospect, with regard to *Astronautics*, the publication of the American Rocket Society. Thus, almost at a stroke, BIS members could expect to receive three publications, instead of only one - albeit one of them was written in German.

The February 1936 issue of the *Journal* was resplendent with a cover photograph attributed to Yerkes Observatory showing eight different views of the planet Mars, the block having been loaned by Eric Frank Russell after its use in his *Ida and Victoria* magazine series. An article by Peter van Dresser (editor of Astronautics) described the 1935 research programme of the American Rocketry Society; there was Cleator's inevitable *Analecta*, a list of ten new members and a somewhat strange and contentious article by Cleator entitled "Weisberger Moon- A Lunar Logogriph".

Analecta included the somewhat startling revelation that according to information received, "the authenticity of which would seem to be beyond doubt", the Air Ministry was at that moment engaged in the construction of an experimental rocket ship capable of carrying men. The outer shell of the vehicle was set to be nearing completion.

Analecta went on to mention that Herr Will Ley in America was planning to launch the world's first liquid fuel rocket airplane and that the American Rocket Society had recently honoured Mr P Cleator by appointing him as an Honorary Member of that Society, "in recognition of his contribution to the science of rocketry".

Cleator's book *Rockets Through Space*, had duly seen publication towards the end of February, and had resulted in the expected avalanche of enquiries concerning membership of the Society. Cleator had been good enough to present me with autographed copies of both the British and the American editions of the book, and in the queue for autographed copies were Arthur Clarke and Edward Carnell. The book itself was a first-class exposition of the problems relating to interplanetary travel and the manner in which these problems might be solved. The book was well illustrated with 22 photographic plates and 21 line drawings Its appearance was well timed by those whose purpose it is to time such things, and its contribution to the progress and well-being of the Society, as well as the interplanetary idea, was immeasurable.

The Annual General Meeting for 1936 had finally been arranged to take place on Friday 20 March 1936, when it was decided that an approach should be made to Professor AM Low D.Sc. with a suggestion that he might honour the Society by accepting a position as Vice President. Professor A.M Low had already accepted an Honorary Fellowship of the Society and had contributed an introduction to Cleator's book.

Professor Archibald Montgomery Low was the editor of the monthly magazine *Armchair Science* and was well known as a writer of popular science, as well as of some science fiction. He had been responsible for over 100 inventions, and amongst his 30 secret patents during the First World War were those for wireless torpedo controlling gear.

In 1914, in London, he had demonstrated and lectured upon the "Low" television system at the Institution of Automobile Engineers, this being followed in 1918 by the accomplishment of infrared photography. During the period 1919 to 1922 he had held the position offered to him by the Army Council of Hon Assistant Professor of Physics at the Royal Artillery College.

Apart from his actual achievements in science, Professor A.M. Low had been prominent when writing of the possible future development of civilisation, and in several of his books and articles he had dealt with the subject of rocket propulsion and interplanetary travel. It is perhaps also of interest to relate that Professor Low's daughter, Ivy, had married Mr M. Litvinov, Soviet Ambassador to the Court of St James, who was later to be nominated as Soviet Foreign Minister.

The Society, of course, already possessed one Vice President, in the form of Colin Askham. Eric Russell, and myself considered that it would be only right if we visited Askham to explain that it was proposed to offer a further post as Vice President to Professor Low. It was not an unusual state of affairs for a Society to possess more than one Vice President, and it had been Russell's impression and my own following our visit to him, that Askham had raised no objection to the idea of the Society having more than one Vice President.

Later, however, when I saw Colin Askham again and mentioned Low's interest in the BIS and the fact that he had accepted a vice presidency, Colin seemed to be extremely annoyed and complained that any Tom,

Dick and Harry could be given administrative position within the Society. The passage of time was to prove, however, that Professor Low, was to become a most valuable member, playing a major role in the establishment of the London branch of the Society, and in the transfer of the headquarters of the Society from Liverpool to London in due course.

I might say that I was very unhappy to find that Askham had not readily accepted the idea of another Vice President, even though he had confessed to me not very much earlier that he had been losing interest in the Society. But an even worse blow was to befall him on the transfer of the headquarters of the Society to London - and I regretted this later event more.

The hope of the publication of a regular monthly magazine on behalf of the BIS was based upon the activities of Mr Ralph Stranger on the Worldwide Radio Research League (WRRL). It is not surprising to find that Ralph Stranger also had associations with science fiction having heard a story called "A Menace from Mars" published in *Wonder Stories*.

His brief but meteoric involvement with the BIS began with his report in the BBC's publication *World Radio*, when in the issue dated 13 September 1935, several paragraphs were devoted to the Society. The purpose of the WRRL was the advancement of shortwave radio communication throughout the world. Reports on radio reception were received regularly from members and were collated and analysed by examiners.

At this stage Ralph Stranger was proposing to break away from his and the WRRL's link to the World Radio and with the BBC, in order to publish a separate journal devoted to the objects of The WRRL and science in general.

In his report in the "World Radio" dated 13 September 1935 Ralph Stranger stated:

"I understand this young Society (the BIS) had difficulties in respect of their organ, which is published in small form, quarterly. Perhaps this difficulty will disappear if we place at their disposal our new journal, especially as our theoretical studies are running in the same channels. Progress in rocket propulsion will interest members of the WRRL."

Accordingly, correspondence ensued between P Cleator and Ralph Stranger with a view to possible cooperation. Cleator mentioned to Stranger, G. Edward Pendray's idea of a possible merger of the journals of the three English speaking societies, with the intention being to produce a combined, well-illustrated, monthly publication, with many more subscribers than if the societies were to continue on their separate ways. The societies involved, of course, apart from the BIS were the American Rocket Society and the Cleveland Rocket Society. The question of cost had killed the idea as far as the BIS were concerned, but the proposed new journal of the WRRL might well be able to take the place of the publications of all three societies. At least, Ralph Stranger would have a fair number of additional subscribers, and the societies would have a combined journal.

Cleator was to write to the USA re the possibilities; in the meantime, Ralph Stranger raised the question of a change of name for the BIS, a proposition which, of course, had already been considered and rejected earlier by the members.

In his report in World Radio dated 11 October 1935, Ralph Stranger announced that with that particular issue, World Radio would cease to be the official organ of the WRRL. This function would be taken over by a new magazine entitled *Ralph Stranger's Science Review*, which was due to appear monthly commencing in November 1935 at a cost of 1s 0d (5p) per copy, or 10s 6d (52½p) per annum.

Secretaries of Radio and other societies who wished to share in the journals should write to Ralph Stranger at once. It was added that the new publication would only be available on a postal subscription and that it would not be available from bookstalls.

In spite of this statement, however, *Science Review* did in fact appear regularly on newsstands throughout the country; after the first three issues had appeared, I wrote to Ralph Stranger to congratulate him on the quality of the publication. At the same time, I enquired as to whether or not a BIS crest could be added to those of the other organisations whose emblems were already appearing upon the front cover of the magazine, and whose representatives were regular contributors.

A reply from Stranger dated 14 February 1936, regretted that Cleator appeared to have dropped the idea of cooperation with Ralph Stranger's *Science Review*, but the offer still stood, and he would be pleased to see a BIS badge and the initials of the Society on the cover of the magazine. He would be delighted to do all he could for the BIS; in the meantime, he requested the loan of the block illustrating the scene from the Ufa film *Girl in the Moon*, that appeared in the October 1935 issue of the Society's Journal.

In spite of the fervent hopes of a regular monthly publication featuring the BIS, three issues of Ralph Stranger's *Science Review* had already been published, without our participation. How this came to pass, I cannot truly say; the arrangements had seemed to have been in the hands of P Cleator, who was, of course, the person most likely to have contributed on behalf of the Society.

Apparently, Cleator had expected Ralph Stranger to keep him informed as to how the arrangements for the publication of his *Science Review* were progressing, and when Stranger had last written to Cleator, it had seemed uncertain as to whether or not the magazine would appear. In addition, during this particular period of time, Cleator was not likely to have been able to spare the time to chase up Stranger, again and to compose contributions to the *Science Review*, as he was fully engaged in preparing for the publication of his book, *Rockets Through Space*.

Whatever may have been the reason for the delay in utilising the services offered by Ralph Stranger, Cleator finally wrote to him on 26 February 1936, stating that he would prepare 1000 words, and would provide free articles for the time being - but in return would like to receive six copies of each issue of the magazine, free of charge, to send to societies abroad.

Indeed, according to Cleator, Stranger would have originally offered payment for BIS contributions, but this offer was later withdrawn.

However, in addition to the free contributions of behalf on behalf of the BIS, Cleator would also like to be able to submit "special" (presumably paid for) articles from time to time. He would also arrange for his publishers to send a free autographed copy of *Rockets Through Space* to Ralph Stranger for review in the magazine.

In view of the arrangements, now being made for the BIS contributions to appear in *Science Review*, I sought the aid of Eric Russell in preparing a small BIS badge to embellish the front cover of the magazine. In due course, I dispatched a sketch to Ralph Stranger, a sketch showing a badge very similar to that used by the BIS at the present time. About half an inch in diameter, the badge showed (in black and white) *The Girl in the Moon* type rocket blasting its way into space. Five white stars were visible with the initials BIS at the base.

It was at this stage in the proceedings, that Eric Russell, and myself decided that it would be an appropriate moment to visit London, both from the viewpoint of interplanetary travel and of science fiction. Wednesday 25 March 1936 was the date fixed for our visit, and Ralph Stranger agreed to see us at 1 pm at his office, which still appeared to have been in Broadcasting House, Portland Place.

We were to arrive at Paddington station at 12.05 pm and amongst other destinations, we had in mind were the editorial offices of the proposed new *British Science Fiction* magazine (the prospective editor having been Mr T Stanhope Sprigg of Newnes Publications).

We were also having what turned out to be a momentous first meeting from the point of view of science fiction in Great Britain, with Ted (Edward John) Carnell and Walter H Gillings, neither of whom up to that time, knew much about each other or about each other's activities. Ted Carnell arranged to meet us at Cannon Street Railway Station, where (he stated) there were several seats outside the barrier of platforms 1 and 2 (where he would rest his weary bones whilst waiting), and he would be wearing a bowler hat and a black overcoat. In order to make it easier for us to identify him, he would be carrying a red-covered book and wearing a red tie to match. All three of us were then to hike ourselves out to the wilds of Ilford to see Walter Gillings.

Ted added that Cannon Street Station could easily be reached from practically any part of London (East or West) by underground either direct to Cannon Street or to Bank Station, which was only 100 yards down a short turning called Walbrook.

In spite of all the detailed instructions about where to meet Ted, when I telephoned him at Gamages Stores, where he worked in the printing department, it was finally decided that we should meet at Liverpool Street Station instead of at Cannon Street. As it turned out, this could have been a most unfortunate change of venue, because Eric Russell and myself, had been waiting for about half an hour before we realised that we had been waiting at Broad Street Station, which was (and still is) the adjoining station to Liverpool Street.

A few years ago, when I had a job as a courier with Littlewood's Pools, I visited the Liverpool Street/Broad Street complex on several occasions, reviving happy memories of my first meeting with Ted Carnell. Fortunately, Ted had not given up waiting by the time Eric Russell and myself were able to wend our way the relatively short walk to Liverpool Street Station.

As it happened, both Carnell and Gillings were to play an important part in the establishment of the London branch of the Society later in 1936 and the transfer to London of the Headquarters of the BIS early in 1937.

Eric Russell and myself were in fact meeting Carnell and Gillings as much on account of our mutual interest in science fiction as from the point of view of the BIS. And just as Phil Cleator had paid for his own air flights to Germany (costing considerably more than our efforts!) Eric and myself were paying our own expenses for the trip to London - 10s. 6d (52½p) return for the day trip. At the end of the day, we left Paddington Station, London at midnight, and after dropping newspapers and milk churns at every station in Wales, reached Lime Street Station, Liverpool, at 8:15 am - just in time for me to go straight to work at the Education Offices.

I sometimes wonder how the course of British science fiction and the BIS might have been affected should we have missed meeting Ted

Cornell at Liverpool Street Station, then proceeding to meet Walter Gillings at Ilford. As it happened, the visit to London turned out to be a monumental success, not only in our encounter with Ralph Stranger and Ted Stanhope Sprigg, but in that the meeting of Carnell, Gillings, Johnson and Russell at that time, turned out to have been a major springboard for British science fiction and science fiction fandom.

On our arrival in London early in the afternoon, Eric Russell and myself also encountered some difficulty in finding our way to Broadcasting House. At one stage, we enquired the direction from a Cockney who was selling papers from a small kiosk; we indicated that we thought that Broadcasting House must be in the very near vicinity. Although our man had obviously been at his pitch for many years, his only comment was, "Oh, I never go in that direction!". Broadcasting House, indeed, turned out to be just around the corner.

We had a pleasant, if a somewhat unproductive interview with Ralph Stranger. My main memory, being his assurance to us that progressive forces were hard at work within the portals of Portland Place, and that the results of such forces would become apparent as time went by.

Following our return to Liverpool, I sent Ralph Stranger a list of BIS members considered to be capable of writing authoritative articles on rocketry and interplanetary travel. At the same time, P Cleator had arranged with Stranger for the first BIS contribution to *Science Review* to appear in the April 1936 issue. This, an official supplement to the *Journal* of the BIS was dubbed *"Ad Astra"*, edited by P Cleator and illustrated by a Test Motor of the American Rocket Society Series 3, Run 1, photographed on the Proving Stand.

In addition to a note about the American rocket motor tests, Cleator submitted a column entitled "Introducing *Ad Astra*", while I, myself, presented a column of secretarial notes; the general tenor of both contributions was that BIS members should purchase *Science Review* every month, and that readers of the magazine, who were not already BIS members would be welcome to join the Society.

At this stage, although we didn't realise it at the time, events were moving to a climax, both in the manner in which the BIS was to be run in the future and our relationships with Ralph Stranger and his *Science Review*.

The cat was thrown amongst the pigeons at a meeting of the BIS Council on April 1936, when it was decided, almost unanimously that I, myself, should become responsible for the production of the *Journal* of the Society, under the control of the Council.

Accordingly, I wrote to Ralph Stranger. On 17 April 1936 to inform him that I was preparing a 16-page Journal, which would include advertisements. I offered him a quarter-page free advertisement for his *Science Review*, if he would let me have a draft of suitable insertion. At the time, I emphasised that all letters intended for the BIS should be sent to Mill Lane, including the six copies of the magazine, which the Society was due to receive each month free of charge.

It was then the Cleator received a long letter from Ralph Stranger complaining about the draft of my secretarial notes, which he had received from Cleator and which were intended for publication. As part of the supplement in the May issue of *Science Review*. Amongst other items, Stranger mentioned that as Hon Secretary of the Society I had failed to write to him offering him Honorary Fellowship, before including in my "notes" the statement that Ralph Stranger "had joined the Society as a Fellow".

Indeed, I had mentioned the matter to him verbally, when we had met at Broadcasting House, but what he was concerned about was that in my notes I should have stated that in view of his valuable services to both the BIS and the WRRL, he had been offered and had accepted an Honorary Fellowship of the Society - rather than that he 'had joined the Society as a Fellow".

The matter having been brought to my attention, I could quite understand Ralph Stranger's feelings about this situation. Unfortunately, this was a method of description that I had copied from P Cleator himself. In his book *Rockets Through Space*, Cleator had referred to certain illustrious foreigners with whom he had made contact, "all of whom had become Fellows of the Society".

This implied that certain personages could hardly contain themselves in their eagerness to join the Society and to pay the prescribed annual subscription. The fact was, of course, that in view of their reputations we were grateful to have their names associated with the Society and to have them as subscription-free Honorary Fellows. However, I can quite see that it sounded better when Cleator wrote it up in the way he did...

Ralph Stranger in his letter went on to state that the secretarial notes submitted to him on my behalf hardly came up to the literary standards to which he aspired in *Science Review*.

This entailed a long letter to Ralph Stranger containing explanations and apologies; my visit to London had left me with a huge backlog of correspondence, at a time when I was being pressed by Cleator to supply my draft of the "secretarial notes".

I had rushed the draft off to Cleator, thinking that he would tidy it up himself if he thought this to be necessary. In fact, at one stage, he had written to me to say that any reports I sent to him for inclusion in *Science Review*, he could always adjust to suit the occasion.

At the Council meeting held the following Wednesday, I mentioned to Cleator that this all too prominent "edited by P Cleator" as a subheading to *Ad Astra* was somewhat in the nature of a slight to myself, as the only other contributor, while not disputing the fact that my rough notes had been edited by the President. I would have preferred something more like "Conducted by P Cleator" or "Compiled by P Cleator".

Be that as it may, the next chapter, opened, when I was informed by Cleator that he had sent in my rough notes exactly as he had received from me.

I hastened to point out in my letter to Stranger that he was proposing to omit my report on the Annual General Meeting 1936, (although my notes, published in the April issue of *Science Review* had promised such a report) and that he had previously given Cleator to understand that we were at liberty to print whatever we decided - even if only of "parochial interest".

As a kind of anti-climax, I went to ask Stranger for a loan of the block of the BIS badge for use on BIS stationery or else a return of our line drawing. In my letter dated 5 May 1936 I wrote to thank Mr Stranger for sending me the block of the badge and enquiring as to whether it was a gift to the Society or whether it was a loan - as it had looked like a new block. At the same time, I asked if permission could be obtained to publish his article, "A Two-Year-Old Mystery" from *World Radio* in the next issue of the *Journal*, which I was then in the course of preparing. Some slight alterations might be required to make the article suitable for presentation to BIS members, and for obvious reasons the article would need to have a revised title of "A Three-Year-Old Mystery".

The article referred to the discovery by Dr Karl Jansky of the Bell Telephone Laboratories, of the USA and which was described in his paper, "Electrical Disturbances of Extra. Terrestrial Origin" published in the Proceedings of the Institute of Radio Engineers of America 21, 1387, 1933.

Dr Jansky, during his study of electrical disturbances and interferences, had discovered a high-pitched hiss of unknown origin the source of which took precisely a year to complete a full circle. The signals were received on a 14.6 metres wavelength (20,548 kilocycles per second) and suggested an unmodulated carrier wave.

Readers of both *World Radio* and the *Journal* of the BIS were asked (were they in a position to do so) to repeat Dr Jansky's experiments, with suitable apparatus and to get in touch with Ralf Stranger at the address of the BBC.

In a letter to Ralph Stranger dated 25 March 1936 (the actual day on which Eric Russell and myself had visited London) P Cleator stated that he trusted that Stranger had regained his normal health, as he had understood that he had not been up to the mark recently.

I had received no reply to my previous four letters to Ralph Stranger when I received a letter dated 6 May 1936 from H.S. Batten (Secretary to Mr Ralph Stranger) to state that Ralph Stranger, who was unable to write to me himself, wished to tell me that the block which was sent was a gift to the Society. Batten added that I was free to make any alterations I may wish to the article "A Two-Year-Old Mystery" but that it was desirable that I should send proof for approval before publication. He went on to say that Mr Stranger hoped to be able to write to me again shortly.

I again wrote to Ralph Stranger on 11 May 1936, thanking him for the gift of the block and enclosing, for his perusal a draft copy of the revised article "A Three-Year-Old Mystery". At the same time, I requested the loan of the block of an astronomical photograph from the January 1936 issue of *Science Review* in order to illustrate the article.

Having introduced in the April issue of *Science Review*, Cleator's contribution intended for publication in the May issue was headed, "Requiem" and detailed the course of events as they had transpired at the Council meeting at which I had been given responsibility for producing the *Journal* of the Society (to which was a supplement). He went on to say "Just how this page will be headed in the next issue is a question to be decided". He then presented me with the argument that was his own title and should not be used in future supplements that may appear in the *Science Review*. He added that the heading "Edited by P Cleator" had been omitted from the May edition. Nevertheless, it still appeared in the supplement, when it was printed in Science Review.

It was my view that no copyright existed in relation to titles, and that in any event, it was a title that had been used by Cleator officially on behalf of the Society. I did not see that Cleator could forbid its use for further supplements. Indeed, it might have seemed somewhat acrimonious to readers of *Science Review* for the title to be changed because Cleator was no longer contributing. According, I sent off number three for inclusion in the June 1936 edition of the magazine.

xxxxx

I then received a letter dated 26 May 1936 from the managing director of Bernard Jones Publications Ltd under cover of a letter from HS Batten. The Managing Director's letter reads as follows : " We regret to inform you that we are obliged to suspend for the time being, the publication of Ralph Stranger's *Science Review*. Mr Stranger has had the misfortune of being thrown by his horse and he advises us that he is now under doctor's orders to refrain from all extra work".

It was added that the suspension must also apply to the work of the WRRL, and that while it was desirable that all observations in hand should be continued, no reports should be sent until so requested. Unexpired portions of any subscriptions to *Science Review* could either be returned or transferred to Television and Shortwave world.

Both Phil Cleator and myself, wrote expressing sympathy and a hope for an early return to full health by Mr Ralph Stranger. In the meantime, I had produced a 24-page edition of the *Journal* of the Society dated June 1936, featuring Stranger's revised article " A Three-Year-Old Mystery" (with no astronomical illustration) and with a stop press announcement concerning the accident, sustained by Mr Stranger.

Fate has an inexorable way of timing events to perfection; and so, it was that the only versions of Ad Astra that saw print were those that had been edited by P Cleator, and that the demise of the supplement coincided with the demise of *Science Review*. I received no reply to my letter of sympathy, and I doubt whether Cleator received any reply to his.

That was over 40 years ago and to this day it is still a mystery to me as to what had actually happened to Ralph Stranger and how he had fared subsequently.

As previously mentioned, the Council of the Society had resolved that I become the new editor of the *Journal*, the next issue of which apart from including paid advertisements- as far as we could solicit them - but would be enlarged to 24 pages. P Cleator agreed to cooperate by supplying me with any information that might be useful to me in performing this task.

Unfortunately, this was not Cleator's only reaction, for I then received a letter from him dated 12 April 1936, in which he tendered his resignation as a member of the Council, adding that he felt sure that matters would proceed more smoothly without his dissenting voice. He sincerely hoped that I would be able to cope with the *Journal* OK but was sorry that he couldn't agree with our plans which savoured of "too many cooks"- though this does not necessarily mean that the resultant broth would be unpalatable to anyone but himself. He did not need to stress the importance of the *Journals* to the Society then added (rather darkly) "by it shall ye be judged".

He went on to say that his decision was due not so much to the changes that the Council had brought about, as to the manner of their being brought into effect. The eight points given in our "ultimatum" had to all intents and purposes, been decided upon by all the Council members but himself hours before the meeting took place. It was obviously not a bit of use him being upon the Council at all. On reflection, I must admit that the position of Cleator must have been very similar to that of King Charles being arrested by Cromwell's Roundheads!

I don't know whether such a course of action occurred to Cleator, but surely it would have been sufficient protest if he had written to say he would not be attending any further meetings of the Council until further notice, instead of resigning from the Council -yet still remain President of the Society.

I replied to Cleator at the same time thanking him for some information he had supplied and enquiring about an article I believed he possessed presumably written by Willy Ley concerning his liquid fuel mail rocket.

In my letter. I also inquired as to whether Cleator wished to remain President of the Society, as it raised rather peculiar constitutional issues to have a president who was no longer a member of the Council. We were willing to arrange a further Council meeting at a time to suit him, when he could state his views. I pointed out that the Council had outvoted a suggestion of mine that the front page of the *Journal* should be used for advertisements. I was content to be the servant of the Council, whereas Cleator wished to be its master and to operate a dictatorship.

I went on to say that I was convinced - and the others agreed with me in this - that if the Society is to make progress, it must be run by the Council, and everything of importance must be passed by that body. Cleator, who was President of the Society and also the Chairman of the Council should advise the Society, keep order at meetings and generally act in a manner commensurate with that office. As Secretary, it was my duty to carry out the instructions of the Council and generally to deal with correspondence and other matters of interest. This included imparting information I may receive to members of the Society, passing on books I may receive to the Librarian, arranging meetings, compiling the *Journal* of the Society, with any aid that may deem necessary, and which it may be possible to obtain.

I added that most of these matters had been attended by Cleator in the past, with the result that it had been difficult to find out what was happening or what arrangements were being made on behalf of the Society as very little information was imparted even to the Council of the Society. Apart from the little that we ourselves gleaned from the pages of the *Journal*.

I went on to say that I would not wish to sever his connection with the Society for which he had put in so much valuable work, that I had only the best interests of the Society at heart, and all my work had been with that end in view. I was willing to bow to the majority decision of the Council, and he himself, more especially as he was the President - should be willing to do likewise.

Had I thought that the discussion that had taken place before the last Council meeting had been unfair, I would not have raised the issues with the members beforehand. As it was, I was taking an important step, and did not wish to take any matters any further, unless they were found to be in agreement with me. I would have liked to have consulted with Cleator but thought that in view of a previous experience that if he knew exactly what was to be proposed, he would have failed to attend the meeting.

In conclusion, I appealed for Cleator's cooperation as his name carried a lot of weight; but the BIS must be run, as a Society, and if he could not agree to this, I for one would be extremely aggrieved.

In reply I received a long letter dated 15 April 1936 from Cleator, in which he went over my own letter point by point. And it was from this time onwards in my relationship with Phil Cleator that no matter which way I turned, I seemed to become confronted by an impasse, all of which, strangely enough, seemed to be genuinely unavoidable in the circumstances.

I had asked Cleator for his list of persons and organisations to which copies of the *Journal* should be sent. This presented an immediate difficulty. There were two lists, but only the President could follow them. There was a list, including members of the Society and those who always received copies; there was also a list of those to whom copies were usually sent, publications newspapers, etc. That might have been called an "optional" list. To add to the confusion, some names have managed to be featured on both lists. I had, of course, the names and addresses of members of the Society, and Cleator would type out a list of other recipients and let me have it. As for newspapers, etc. it was up to me to decide for myself, which should receive the copies of the *Journal* and which should not. A further complication was that certain distinguished members abroad, were in the habit of receiving more than one copy.

Cleator went on to say that the articles about Will Ley's experiments, did not exist; he had merely proposed to write an article on the subject himself from information received from Ley. He had given the experiments a brief mention in, so it may be that no more needed to be said about that matter. Indeed, running such an article in the *Journal* would be a waste of space as Ley was shortly to publish an account of those experiments in *Armchair Science*. Cleator added that he had not known about the proposed article in *Armchair Science* when he had planned to give an account in the publication of the Society, and in any event, by the time the next issue, the *Journal* appeared the news would be more than a little stale.

Cleator then went on to discuss the question of unpaid advertisements which by tradition he had included in previous issues of the *Journal*. He had promised Mr Chambers that he would give a brief mention of Chambers' Encyclopaedia, a new edition of which has just been issued - this in return for the Society and Cleator's book being publicised in the forthcoming May issue of Chambers' Journal. Cleator stressed that this was the most unusual cause for so esteemed a publication as Chambers Journal and no mean compliment to the BIS. Cleator would write the usual Chambers' Dictionary note for *Analecta* - if I intended to retain such a column.

In view of the fact that we hoped to be able to attract other advertisers, the advertisements, apparently by Chambers, could do some good. In addition, a note might be included about the proposed American Edition of *Rockets Through Space* and he had promised the British publishers, Messrs Allen and Unwin, free use of the back page of the *Journal* in order to advertise his book further.

Cleator went on to say that he had anticipated the difficulty with regard to the President not being a member of the Council. He would be delighted to be able to retain the Presidency but how this was to be managed was a matter for the Council themselves to decide. Cleator then surprised me by saying that he was going away. He did in fact explain to me where he was going - which was a private, personal and confidential matter. So, it appeared that the action taken by the Council had indeed been taken at the most providential moment, in spite of everything.

Cleator ended his letter by saying that the root of the trouble seemed to be that he put more faith in dictatorships than in Councils, with no reflection on the Society's Council. He could only quote "too many cooks" again. It was far easier for one person to decide to do a thing and do it than it was for a body of persons. Hence, with such heretical views, the Council was no place for him! He was more than prepared to abide by the decisions of the Council, and just what his exact position would be in the future, he left the Council to decide. He concluded by assuring me that there was not the slightest feeling of ill will at all on his part.

Unfortunately, I got into further hot water by conveying to Cleator the decisions made by the Council at their meeting on 24 April 1936, when his resignation from the Council was accepted "with regret". A standing invitation was extended to him to join in their discussions, if in the future at any time he decided to associate himself with the Council again. Until the next Annual General Meeting (when officers and members of the Council would again be up for re-election by the membership) he would continue to be regarded as President of the Society.

In his letter dated 28 April 1936, Cleator took me up on my statement that he would continue to be "regarded" as President of the Society. He thought that the position was rather ambiguous. "Who", he asked, "as the word 'regarded' suggests was the President in actuality?" He had not the slightest objection to being the President, but resolutely refused merely (as a sort of favour, perhaps) to be regarded as President. Would I clear up this point at my convenience?

Cleator had certainly shot me down in flames and I hastened to assure him that he had clothed the word in meanings far from those intended. Indeed, if it were possible for things to be so arranged in accordance with the constitution, I added that doubtless, he could be made permanent President, should he so desire. I went on to emphasise that I would willingly meet him at a mutually agreed time and place in order to discuss our differences. Members of the Society outside Merseyside would be bound to be bemused at the idea of a president no longer a member of the Council of the Society.

I am reminded of a similar incident in the Liverpool Education Offices wherein the then Director of Education (who apparently decided to abolish Deputies to the Heads of sections) ruled that I had been appointed to "act" as a Deputy to the Head of the School Meals section, I was not in fact, the Deputy to the Head of that section! I was only "acting" as the Deputy.

Further difficulties arose, inasmuch as having run out of supplies of Society notepaper, I found that Cleator did not have any available that he could let me have. Additionally, I did not feel that I was receiving carbon copies of all the letters from Cleator that I might have expected to receive- but he explained, not all the letters he sent were officially on behalf of the Society- others were private letters.

Despite the protracted negotiations between Cleator and the Council (of which I was invariably the middle) I somehow managed to maintain my correspondence with Arthur Clarke, Edward Carnell, Walter Gillings, Francis Knight and in particular- J Edwards. Following our visit to London, Eric Russell seemed to have disappeared into a limbo, probably induced by the writing of science fiction.

In the course of my correspondence, a rather strange comment that I made to Gillings was that Russell (of all people!) did not think that science fiction and rocketry went hand in hand, whilst it was my own view that the two were virtually inseparable. Gillings, on the other hand, thought that "all this energy" was wasted upon rocketry and would be better expended in the cause of science fiction…

Walter Gillings was at the time planning to edit and publish his science fiction fan magazine, which eventually appeared as *Scientifiction, The British Fantasy Review*. Having collected about 1,000 addresses in my capacity as Hon Secretary of the BIS, I somehow found time to type out that number of envelopes for Gillings to circularise those who might become interested in subscribing to his proposed magazine. The circulars also included an appeal for members of the British Interplanetary Society.

Most of Arthur Clarke's numerous interesting letters to me about this period of time nearly always had to be acknowledged by a postcard. He now also seemed to have become involved in the midst of voluminous correspondence with Eric Russell and was even able to claim 10% participation in one of the stories Eric had accepted for publication in America.

Arthur went on in his letters to discuss various technical matters of interest, such as the greatest height an aeroplane might be able to ascend, the design of powder rockets (he knew very little about ordinary rockets), the velocity of light, the size of the Earth's orbit and whether radium was a compound.

Having heard of the BIS situation concerning Phil Cleator, Clarke stated (prophetically) that he couldn't take over from him yet- "Perhaps later, who knows?". He went on to inform me that he had finished 26th out of 1500 in the Civil Service examination. Which was better than I had managed in 1930, which was 552 out of 2500. The first 500 were offered posts...

They say that the hour produces the man- and the man of the hour undoubtedly turned out to be J Happian Edwards. I was preparing for my first issue of the *Journal* to appear directly under the aegis of the BIS Council, and seeking paid advertisement to assist with the cost of production. Edwards, then being associated with the 362 Radio Valve Company was approached in this respect, and I was pleased to be informed that his company was prepared to advertise as requested. Their paid advertisement did in fact (at Edwards' special request) appear in the form of an article in the *Journal* headed "Some problems in the Navigation of Spaceships".

Accompanying his letter agreeing to place the advertisement in the *Journal* Edwards sent a separate letter dated 22 April 1936, which I submitted to the Council of the Society for consideration. The Council decided that this letter, together with suitable editorial comment should be published in the forthcoming (June 1936) issue of the *Journal*.

The letter from Edwards, and my own comment, read as follows:

Dear Sir,

A London Section?

I've been making a few contacts, since my last letter, and though a considerable amount of useful work may be initiated by putting individual members in touch with one another a lot more would result from a London meeting, no matter how small. I am sure that the members would be glad to meet somewhere at their own expense, even if unofficially. This could easily be arranged, and I would be glad to do any arranging necessary.

Members could be communicated with individually, and accommodation was arranged for as many as were intending to come. By making it unofficial everyone would understand that it was at our own expense.

For other reasons, however, I think, a permanent London branch is advisable if only in the form of a local secretary who would collect subscriptions etc. as I think this would appreciably increase the membership.

I have been doing my best to get new members (whilst limiting my activity to such as are likely to prove useful). I find that while it is easy to interest anyone and make them agree that they will join, it is another matter to get them to say, "I will send off my subscription **now**." If there was someone available to take such subscriptions in London, it would be fairly easy to clinch the matter.

Further, with regards to finance, it is obvious that large sums will be needed for research and that it behoves any members who might be able to invest any money to do so. But I feel he will be severely handicapped on asking someone to dub up a thousand pounds or so (with little hope of any advantage from so doing) if he uses the "Members notepaper". I can't help feeling that it looks more "Amazing storyish" than like a serious scientific society.

Hoping that you will not consider that I am being unduly critical for a new member.

I remain,

Yours faithfully,

J H Edwards.

C/o The 362 Radio Valve Co Ltd

Stoneham Works

Stoneham Road

Northwold Road

Upper Clapton E.5.

The editorial comment was as follows:

> If sufficient support is forthcoming and satisfactory arrangements can be made, the Council will consider the formation of a London Section. Those interested in an informal meeting such as is proposed should communicate with Mr Edwards when the necessary arrangements can be made. Mr Edwards, in deference to his own request and his enthusiastic work in stimulating the interest of Londoners in the Society has been appointed our official representative in the Metropolitan area. We agree with the remarks concerning the notepaper, and at a recent Council meeting, it was decided to adopt a new, more dignified design for member stationery. This will be issued when present low stocks are exhausted.

Edwards had been very pleased to take on the job of local representative (pro tem) and would do his best to make London a worthwhile proposition. When he saw the new members' notepaper, he thought it would be a big improvement, and the inclusion of Ralph Stranger amongst the Fellows would be very helpful in obtaining new members as he was well thought of in London.

My version of the *Journal*, duly appeared dated June 1936 of 24 pages, illustrated, and listing no less than 24 new members, no doubt all from the catchment area of *Rockets Through Space*. The issue boasted no less than five photographic plates, including a small photograph of Colin Askham and a line drawing of Professor Low.

Arthur Clarke wrote enthusiastically saying, "Congratulations! A really hot *Journal*. You have made a really fine job of it - in fact, it's more like a magazine than a journal now".

Unfortunately, Edwards wrote at the end of July to say that the only response he had received from the publication of his letter in the *Journal* turned out to be enquiries concerning the health of Ralph Stranger!

On a further visit to London, I arranged to meet Edwards at Euston Station and here again we were unlucky in missing each other, complicated by the fact that we had not previously met, and as there was no definite barrier to number one platform where we had arranged to meet - and Edwards hadn't realised that there would be so many people about!

In spite of the apparent lack of progress at this stage, from this point onwards events moved swiftly, and when "That lot up there" (whoever they are) decide to move things - they move! It was even an appropriate moment for Arthur Clarke to go and live in London - "in the service of the state". He had been appointed to the Board of Education "So we might clash", he said, referring, of course to the fact of my own employment, with the Liverpool Education Committee! He hasten to get into touch with J H Edwards. At this time, Clarke has been doing a great deal of writing and Eric Russell had commented favourably upon his style. "So I might go on to better things", Arthur prophesised.

To add to the sudden impetus, I was able to send to Edwards, a list of BIS members whose addresses were such that they would be able to attend meetings in the metropolis. Furthermore, Eric Russell was preparing to visit London in the course of business. He had arranged to meet Carnell, Clarke, Edwards and Gillings, as well as T Stanhope Sprigg.

Following the publication of the letter from John Russell Fearn in the March 1931 issue of *Amazing Stories*, I had visited John and his mother on numerous occasions at a home near Blackpool. I even went so far as to collaborate in the writing of a time travel novelette called originally *Amen*. With the extension of John's writing activities and the obvious difficulties of collaboration with one of us living in Blackpool and the other in Liverpool, John had renounced his association with *Amen*. The story was then re-written in collaboration with Eric Frank Russell. Russell on his visit to Sprigg was hoping to interest him in the re-written story (now called *Seeker of Tomorrow*) for publication in Newnes' projected British Science Fiction magazine.

In the meantime, Clarke had passed his higher Oxford, which he thought might be useful if he went for his BSc. He wanted to get into the BIS swim in a big way. So our positions could be reversed and I could write to him, instead of him writing to me!

Gillings, having contacted Edwards reported that Edwards had "great schemes" but still insisted that all this work was being wasted on the BIS and that rocketry could follow science-fiction.

The BIS had started to prosper, but - to put it in Cleator's terms - I, myself, was preparing to blow up. I was not only Hon Secretary of the Society, I was also Hon Treasurer and editor of the *Journal*. Members were pouring in (thanks of course to Cleator's book) and the formation of a London branch was in prospect.

Eric Russell and G Wilkinson (member - Bootle) gave what secretarial and other assistance that they could, but it was obvious that this could only be a temporary expedient. With the situation such as it was, with the enthusiasm of JH Edwards and the gathering of a nucleus of members in London there could only be one solution: a transfer of the BIS headquarters to the Metropolis.

To have nurtured an interest in science fiction and interplanetary travel during the period of 1933 to 1937 was to have been like "one whose voice was crying in the wilderness". It was unusual to be aware of even one other kindred spirit. The visit, therefore, in March 1936 of Eric Russell and myself to London to see Carnell and Gillings, left Carnell in particular in a state of euphoria. "Really have a lot to thank you for," he wrote in June 1936, "except for an invitation to meet you and Eric Russell, I would not have met Walter Gillings and become so interested in his (proposed SF) magazine".

Ted Carnell went on to say that he went to tea every third Friday to Gillings' place and liked him immensely as a friend. One of Gillings' earliest SF contacts, Len Kippen, would also be present, and Carnell was also in touch with Donald Wollheim and Julius Schwartz, both to become well-known names in the world of American SF. "I now have a very interesting circle of friends", Carnell went on. "Most thanks are due to you, I believe. ...."

Ted Carnell had great praise for Cleator's book *Rockets through Space*-Grand effort, both from the fans' point of view and also the man in the street, who does not know too much about rocketry". Ted went on to discuss the problems of weightlessness in space and the proposed use in literature of magnetised shoes envisaged both by Cleator and John Beynon (Harris)to enable spaceman to walk inside the metallic walls of spaceships. Does magnetism work in a vacuum? He supposed, it must. I hasten to assure him that magnetism would indeed apply in a vacuum, and that it was a kind of "etheric strain" similar to gravitational and electrical fields......

On 6 July 1936 Ted Carnell joined the BIS, adding that he could not say whether he could lend any valuable help to Edwards, but perhaps later on he could meet him and attend some of his proposed meetings. By the middle of August, Carnell had met Edwards, "despite being busy getting a new dance band together for a number of gigs and weddings, that he had in hand".

Ted waxed lyrical over Edwards' character and capabilities. "About 36, but I was quite agreeably surprised in finding him so youthful. He is a red hot, rabid, enthusiastic, rocketry fanatic and I don't mean that usual type of SF fan. Edwards is totally different from any other men I have ever met, and what a brain the man has! He's a mathematical genius, rattles off gigantic equations, computes speeds and heats and other essentials to astronautics just like I would read a newspaper. Has the most widely varied and colourful scheme for putting the BIS right in the forefront of all the countries in the Earth".

"Put forward the idea of a suicide club, in which a spaceship would only have to be fuelled for a one-way trip - there wouldn't be any return - thus saving a huge sum of money, the occupants (members of the suicide club!) heliophoning their adventures and experiences back to Earth, sometime before they eventually meet their deaths by one means or another in outer space".

"Edwards reckons that atomic energy will arrive before very long, cheap enough to use for propulsion, but even with a thick lead shield around the combustion chambers, it would be almost impossible without some super cooling system, to be able to survive within such a spaceship". (N.B. this was 1936!)

Carnell continued "Edwards has been working for some weeks with a friend, trying to work out the speed at which the sun's rotation is slowing down. He says it must have been travelling fast enough to throw off the planets of our system, and as it is no longer doing that thing, ergo, it must be going slower. Then as it is losing mass every moment, it must be getting lighter, which should tend to make it rotate faster. As it couldn't be slowed up by tidal friction then what is the cause for it?".

"Edwards claimed that he could build a time machine now if he wanted to the only snag being that he could only get it to go forward in time, not backward, and that it would be impossible to get back to the present time, once the journey had been started."

Carnell went on to write, "Whatever conclusion you may draw from these words, believe me, the BIS couldn't be in better hands if you wanted it to progress. Edwards has all the right ideas and providing there is a good deal of support coming from London, then he'll make a better go of it than has ever been done before. He certainly won't hear of the BIS packing up".

On 21 August 1936, in a full page, closely typed letter to Carnell, I stated "I would like to see little better than that the Society should be built into something really worthwhile by the London fans. It has no real future on Merseyside (not for a long time, anyway - longer than I can shoulder the work single handed) and I would think that my work would not have been wasted if it should go ahead in London."

I went on "my idea of the situation is that we should circularise members telling them that the Society is being transferred to London - giving the London address and asking for the general consent of the membership to ratify this change. I will transfer the matter to yourself and Edwards gradually until the end of the year. You could then hold the Annual General Meeting in London, at which some of the Liverpool members will try to be present." We then had about 100 members of which we still had about 20 Honorary Fellows.

Progress towards the proposed transfer of the BIS to London, had been slowing down during the late summer of 1936, as J H Edwards, had been busy at work on the Radio Show at Olympia. It was there that Carnell was able to meet him in September, when they were joined by Eric Russell on his business trip, as well as Gillings and Clarke. Also visitors to Edwards at Radio Olympia, were Maurice Hanson and Dennis Jacques of Leicester and Nuneaton respectively. The two last names were joint editors of the *Nova Terrae* (*New Worlds*), a science fiction fan magazine, which they produced on behalf of Chapter number 22 of the Science Fiction League.

During the course of his correspondence with me, Carnell posed various questions on behalf of Edwards: Who is on the BIS Council? What does the Council do? Who orders things to be done? What happens when the BIS is transferred?

Carnell's part in the future of the Society depended upon how much (or perhaps how little) typing needed to be done, but Edwards thought there should be enough members in the London area to be able to undertake the work. Transfer of the Society would depend upon how many London members rallied around. With regard to the publication of the *Journal* (the next issue of which had been delayed by the perennial cause - lack of funds), Edwards thought that it was its production was best left in my own hands for the time being.

The amount of funds available for the production of the *Journal* was further aggravated by the little matter of £3-0-0 to be repaid when "circumstances allowed". 'Pay it into the Research Fund when you can scrape it together", Colin Askham had said, and it was felt that we ought to be able to resolve this matter before publishing the new issue of the *Journal* or transferring the Society to London.

In the meantime, I received a letter dated 14 September 1936 from P Cleator which read as follows:

> "Confirming our telephone conversation of today regarding the transferring of the Society to London. Should this course be definitely decided upon at the coming meeting on Friday next, or at any future meeting, I wish to tender my resignation as President, the resignation to take place immediately after the decision is passed by the members of the Council.
>
> I shall be glad to hear from you in due course."

There was a covering letter accompanying the letter of resignation, which said:

> "I wish this, because, after all, the matter insofar as I am concerned is the subject of a protest. But please do not misunderstand my attitude. I am quite prepared to concede that the Society may progress better if directed from London, and I naturally wish it to progress as much as possible. So much, indeed, that I do not wish to retard progress by remaining President. Far better for the entire direction to be situated in one spot."

On 19 September I replied to Cleator, as follows:

"At the meeting of the Council last evening, it was decided that it was very desirable that the London section should be formed as soon as possible, and that should they so desire it, they should take over the management and property of the Society; the transfer to be confirmed by the membership at the Annual General Meeting in 1937.

Accordingly, it was decided to accept your resignation as President of the Society as suggested in your letter of the 14th instant".

Various meetings then took place between the principals involved in London, including a three and a half hour meeting at Lyons in the Strand between Carnell, Clark and Edwards, who had written to Professor AM Low to try to gauge the extent of his interest. Following a meeting between Edwards and Walter Gillings with Professor Low, the professor agreed to put the outer office of "Armchair Science" at the disposal of the London members for a meeting, and furthermore promised that he would attend himself.

As Ted Carnell put it: 'Joe Lyons little knew that one of his small bun shops in the Strand would be used for the forming of great plans involving the outer reaches of space....."

The upshot was that Tuesday 27 October 1936 was chosen for the first meeting in London of BIS members and anyone else who happened to be interested in interplanetary travel; Carnell and Edwards drafted a suitable letter, which was sent out dated 19 October 1936. The meeting commenced at 8 pm on the evening concerned; 17 members and five friends attended. Professor Low's office was "jammed tight" and Carnell dubbed the gathering, "An overwhelming, enthusiastic success!" This was especially gratifying, as a list that I had sent to Edwards of persons who had made enquiries about the Society did not arrive in time for them to be notified.

Professor Low mentioned Lord Nuffield and AV Roe both great friends of his, and whom he said he could attract to the Society if need be. Carnell suggested that Low might become President of the Society if Cleator was no longer interested, but Low thought things would have to be made very clear to Cleator beforehand as Low was in a very awkward position, knowing Cleator so well.

As it happened, there would have been no difficulty with Cleator over this as his letter to me dated 23 September 1936, he had actually advocated the appointment of a President from amongst the London membership "providing plans to move the Society south materialise".

Edwards was concerned about Askham's position, and if he could still be Vice President if the Society's headquarters were moved to London,

But Carnell suggested that there could be no idea of a transfer until the London branch had been made into a worthwhile proposition. There would be close cooperation with Liverpool and complete autonomy locally, but no thought of taking over at this time.

A small committee was formed with a quorum of 3/7 including Edwards, Clarke, Strong, Carnell and KW Chapman, who met at a little Italian chop suey joint off Piccadilly, to which they had adjourned after the meeting that Professor Low's office. Edwards had received enthusiastic letters from members who attended the meeting, and as "all are raving mad" (as Carnell put it) about the whole idea of interplanetary travel and the BIS there must be another meeting shortly.

At the inaugural meeting, group photographs were taken and forms autographed as souvenirs, and already suggestions were being made for various activities, such as a visit to Greenwich Observatory, lectures during the forthcoming winter and films to be shown at a small cinema. Edwards' fiancée, Miss Huggett, would handle all London and Southern correspondence, at least until an Annual General Meeting could be held.

At the meeting, Carnell and Edwards, had very wisely played down the idea of the transfer of headquarters of the BIS to London, as this would obviously have been too large a pill to swallow at the first gulp. In addition, there were many questions to be answered and problems to be solved before the transfer could take place-this, in spite of my insistence that Liverpool could not hold out much longer - mainly in the form of myself!

One of the main misgivings in the minds of London members was the constitutional aspect of the proposed transfer: who was on the Council of the Society? Was my own position backed by Council? What was the position with regards to Cleator, Askham and Professor Low as Ex-President and Vice president respectively.

I explained to Edwards, that the Council had been constitutionally appointed at the Annual General Meeting in 1936, when the whole membership of the Society had been given the opportunity to attend. The information and directions that I had conveyed to Edwards and Carnell were following upon Council Resolutions, and any action that had been taken by me was subsequently reported back to the Council at the following meeting.

Members in Liverpool would not object to the idea of a transfer (apart from Cleator and possibly Askham) but would in fact welcome the suggestion, for as things were at the time, the management of the Society was in a hopeless state. Appearances had been maintained, but we had lost support at the meetings. Following a full attendance at the Council meeting held on 10 July 1936, the most we had been able to gather together, had been four or five members.

Cleator's resignation from the Council had reduced the number of members from six to five but by the end of 1936 - by constitutional means - the number of members on the Council had been expanded to 11, including Professor Low, who, of course would not normally be expected to attend by travelling from London. The remaining members were Colin Askham, A Beardwood, W Dunbar, R Eddon, Hodgson, L Johnson, T McNab, E Russell, N Weedall, and G Wilkinson.

In the meantime, Walter Gillings, in his capacity as news-hawk had been able to attend at the inaugural meeting in London, and despite the numbers present was not optimistic about future prospects. (At this point, in typing my narrative, I discovered to my sorrow, that

Walter Gillings had died suddenly on Tuesday 17 July 1979, following a heart attack; he was 67 years of age).

According to Walter Gillings in his comment on the meeting, Carnell and Edwards seemed to have no idea how to conduct a meeting - and according to Carnell, Edwards had no idea how to conduct a meeting! Gillings, however, was able to achieve a successful interview with Professor Low and was responsible for a considerable amount of welcome publicity in the national newspapers.

In his correspondence with me, Walter went on to say that those present at the meeting had little idea of what they were going to do or how to do it, various doubtful appointments were made or not made, and he concluded that although it was of no concern of his, he wished them all well.

On the other hand, Wally wished that he could have promoted science fiction in London in the same way and as it turned out, this was exactly what had been doing, without realising it. Indeed, and in spite of himself, Gillings had so much to say that was worth saying, that by the end of November 1936, he had joined the BIS as a member and was promptly voted into the new General Committee.

By early November, matters had drifted to an extremely low ebb in Liverpool, and only Eric Russell and myself seemed to be either willing or able to take a hand in the work of the Society. The remaining members were either indulging in night studies or were engaged in other business that prevented them from attending meetings. The Council was barely in existence, and it was difficult to obtain even a 50% attendance at Council meetings.

I was acting as Secretary, Treasurer, arranging meetings, was responsible for the publication of the *Journal*, and for doing 1,001 other things that needed to be done. It seemed that the conflict with Cleator having been resolved with his resignation as President and Council member, the Liverpool members of the Society had experienced a form of anti-climax and had lost interest in the BIS.

If Phil Cleator, at that moment, had had a kind of wry smile on his face at the course of events, one could hardly have blamed him.

It had been made clear at the Inaugural Meeting of the London branch that a great deal of confidence would be created if one or more members from Liverpool could attend a meeting in London and deal with the many questions that remained to be clarified.

Accordingly, I arranged that my fiancée (Miss Hilda Crossen) and myself would visit London on Sunday 15 November 1936, in order to give London members an insight into BIS affairs. We would arrive at Euston Station about 2 pm, meet the London committee members in the afternoon, then proceed to the Mason's Arms, Maddox Street, for 7 pm where the Londoners would congregate.

Unfortunately, at this time, Ted Carnell's state of euphoria dissolved into one of exasperation. In response to my pressure for an early transfer of the BIS to London, he finally had an outburst: "How could Edwards takeover in view of the poor early response? It would have scared the meeting if an immediate take over had been suggested. They were all willing and eager to shoot rockets, but it was different when it came to routine work." Carnell was answering queries, arranging meetings, printing notices etc. The BIS had already cost him a small fortune. But his final outburst really set me back: "I'm the Les Johnson of London," he had blurted out in his letter of 9 November 1936.

Having blown his top, Ted went on to say that Edwards had divided the secretarial work between Miss Huggett and KW Chapman. Edwards then suggested that London members should pay 6d (2½p) per week in addition to the normal subscription, and that he and Carnell should pay an additional 5s0d (25p) per week each. With the resultant total of £1-0-0 per week, members could be provided with a regular club room, which would always be at their disposal. Ted went on to say that he was "not going to be a mug", then added that he had enjoyed the donkey work up to then - and that he expected that he would continue to do so.

A letter that I had received from Edwards intimated that a transfer was not as far off as Carnell had implied. I presumed to encourage Ted in his endeavours, advised him not to let any one person dominate the London scene - and from the practical point of view, to carry a small notebook in which to jot down any expenses he might incur that could reasonably be laid at the door of the Society. I also hastened to supply him with adequate quantities of notepaper, envelopes, copies of the *Journal* and other literature pertaining to the Society, and printing blocks.

Walter Gillings and R Smith had been added to the London committee to make up the seven required, but when a new constitution had been adopted at the Annual General Meeting, this committee would be disbanded, while others would probably be formed for special purposes.

Over 30 persons had been present at a recent meeting held in London. The majority not having been present at the first meeting - so "Heaven help us" said Carnell, if they all turned up at the same meeting. Carnell had been appointed Director of Publicity, with a Publicity Council consisting of himself, Gillings and J Strong, plus any other member who may have had the experience of writing.

It is clear that in spite of the enthusiasm and wholehearted cooperation given by J Edwards and Miss Elizabeth Huggett, it was to Ted Carnell eventually that the BIS (and indeed I, myself) owed most in the early days of the London branch, and immediately following the eventual transfer of the Society.

In the meantime, Hilda Crossen and myself had duly journeyed to London on Sunday, 15 November, and had met members of the London General Committee and attended the meeting, arranged at the Mason's Arms. We had taken with us the Minute Book of the Society and copies of the three different versions of the constitution that had thus far been tried out by the Society.

In order to keep Hilda company, Rene Cloke (later to become Mrs EdwardsJohn Carnell) had attended; the two girls got on well together until they started discussing clothes (as women will!) As Hilda related to me - Rene, admiring Hilda's all green outfit remarked, "but I can't stand coloured shoes". Her gaze then travelled down to Hilda's shoes-also coloured green - and it was Rene's turn to colour! The two girls then dissolved into a bout of embarrassed laughter that cleared the air a little.

In retrospect, it is very often the silly little trivialities that remain with us in memory when more important items have been forgotten. So as the gathering (predominantly male, of course) was emerging from the Mason's Arms at the end of the momentous meeting, two ladies of obvious easy virtue sidled towards them- but retreated somewhat abashed on seeing Hilda and Rene leaving the hotel with the men. They must have thought their pitch had been queered…

One of my own most persistent memories after the meeting was of Hilda and myself dashing to catch the late train back to Liverpool, accompanied by a very attentive Arthur Clarke. I was only 22 years of age at the time, and Arthur must have been about 19; describing Hilda as a "martyr to science", he was so kind and considerate to us both that it became almost embarrassing. This was especially so, as we knew he was going in totally the wrong direction in seeing us to the station, and he seemed to have been completely unconcerned as to how he could make his way back to his digs. Hilda and myself arrived home at 4.45 am (I, myself, feeling a little sick after all the excitement and travelling) while Arthur arrived home at 2 am. He could then have pondered upon the fact that at the meeting he had been appointed Hon Treasurer at the London end of the Society.

At this point, I can do little better than to quote the report on "The Proceedings of the London Branch of the Society", which I edited and published in the February 1937 issue of the *Journal*, and which cover the meetings of the London Group held on Tuesday 27 October 1936, Tuesday 10 November and Sunday, 15 November 1936 as follows:

Inaugural Meeting

The first meeting of the London members of the BIS to place on Tuesday 27 October 1936 at Professor A Low's offices, 8 Waterloo Place, Piccadilly and resulted in the formation of the

London branch of the British Interplanetary Society. Success was assured from the outset by the enthusiasm with which all present greeted the formation of the branch, and the subsequent difficulty in keeping rigidly to the programme was somewhat allayed by the friendly spirit in which Professor Low, who was present, endeavoured to make everyone feel at ease.

Mr J Edwards who suggested the formation of a London branch in the last issue of the *Journal*, opened the proceedings with a few remarks about the vital significance of the occasion and proposed Mr E Carnell as Chairman. The motion was seconded by Professor Low and carried by the meeting.

Mr Carnell gave a brief summary of the Society's activities during the past three years, drawing special attention to the financial position. It was obvious, he said, that the new schemes would have to be tried to attract more members.

From this he went on to describe events leading up to the resignation of Mr P Cleator, former President. In May 1936, he explained a Council had been formed in Liverpool to consider and approve all actions on behalf of the Society. Mr L Johnson of Liverpool, who had been Hon Secretary of the Society since its formation, had become dangerously overworked and as a consequence, in June, the idea of a London branch was first seriously formulated.

If successfully founded, this branch will eventually become the Headquarters of the Society. Mr P Cleator, the President, disagreed with the proposal that the Headquarters should be moved, and as the movement was strongly supported, he had resigned from the Presidency. However, he promised that the Society would still have his every support. The Society, continued Mr Carnell, would always be grateful to him for the valuable assistance he had given in helping it to its present position.

Speaking at the proposed transfer of the Headquarters to London, Mr Carnell said there could be no thought of any transfer until such time as the new London branch had justified itself as an important unit of the Society. This, it was hoped, would eventually come to pass, as there was a larger number of members in London, than anywhere else. One could also look forward with confidence to the time when experiments could be conducted by the London membership, including as it does many technical experts.

The election of Branch Officers resulted as follows:

President Professor A Low, DSc; Joint Hon Secretaries, KW Chapman and Miss E Huggett; Hon Treasurer, A Clarke; Director of Research, J Edwards; Director of Publicity, E Carnell.

Professor A Low gave an extremely interesting address, drawing a comparison of prejudiced public opinion throughout the ages with the aims of the Society today, and emphasised the importance of not being discouraged by this, or, in our turn, becoming prejudiced. He touched lightly upon the possibility of life forms, totally unfamiliar to man, existing on other planets and even upon spiritualism, which, he thought, we were hardly in a position to accept as yet.

He offered interesting suggestions for increasing the present membership (which were held over for consideration at a later meeting) and concluded by asking if the London meeting had the full approval of Liverpool. In answer, the Chairman declared that he had been in constant communication with Hon General Secretary in Liverpool to this effect, and that the meeting had the good wishes of the Council of the Society.

Mr J Edwards gave a short address in which he mentioned the enormous amount of work waiting to be done in the way of research and investigation, and pointed out the wide fields these would have to cover. He outlined his suggestions for tackling outstanding problems, and requested the wholehearted cooperation of the members in an endeavour to reach their solution. All contemporary science fiction authors he continued, but the ultimate time for space travel of the year 2000, but he was firmly convinced that it would become fact before that date. He foresaw consternation amongst the governments of the world should any particular country succeed in the conquest of space, and thought the only way out of any complications would be a united International Astronautical Society composed of all the societies in the world.

A general discussion ensued on the advisability of changing the name of the Society to one less imaginative, but no decision was made. A number of other important questions were referred for consideration at the next meeting of the branch.

Professor Low proposed a vote of thanks to the Chairman, and the members passed a vote of thanks to the Professor for his kindness in allowing the meeting to take place at his offices, and for his enthusiastic support of the new branch. Mr L. Klemantaski took a number of group photographs of the gathering, which it is hoped will mark a great step forward in the progress of the Society.

Several reports of the meeting appeared in the press, during the remainder of the week.

An informal meeting

An informal meeting of the London Branch was called on Tuesday 10 November 1936 with the object of promoting general discussion, so that the views, interests, capabilities and intentions of the members and other interested persons could be ascertained. A strong feeling developed that it would be necessary to overhaul the Constitution of the Society before the London branch could satisfactorily carry out its work. It was suggested that a Constitution Committee should be set up to consider this point, and report back to the London membership. As the meeting was informal the committee could not be officially appointed, but in view of the fact that Mr L Johnson was coming from Liverpool on Sunday, the suggested members of the committee would commence their work immediately. A meeting was to be called in honour of Mr Johnson, on the Sunday evening, when the committee could be officially appointed.

A Visit from Mr L J Johnson

A meeting was held on Sunday15 November 1936 at 7 pm at the Mason's Arms, Maddox Street in honour of Mr L Johnson of Liverpool, who has been the Hon Secretary of the BIS since it was founded in 1933.

Mr E Carnell was elected Chairman of the meeting. Mr Carnell introduced Mr Johnson to the members, and explained how enthusiastically and unselfishly he had worked for the Society. In reply, Mr Johnson gave a summary of the history of the Society, reading extracts from the minute book from time to time, and giving particular attention to the struggles of the Council to draft and put into force a Constitution which would be satisfactory to all concerned. He said he thought the Liverpool members would be glad to cooperate with the London members in this matter.

Mr R Smith proposed that a committee should be set up to draft a constitution for submission to the membership. Mr Carnell seconded the proposal, which was carried unanimously. It was decided, also unanimously, that the members of the committee should be as follows: Messrs J Edwards, E Carnell, A Clarke, K Chapman, RA Smith, W Gillings and Miss Huggett.

A vote of thanks was then accorded to Mr Johnson in appreciation of the trouble he had taken in coming to London to address the London members.

Following my visit to London. I wrote to Edwardson on 17 November 1936, Enclosing the latest draft Constitution of the Society, which Edwards found nearer to the views of the Londoners than the previous versions he had received. I pointed out to Edwards that it was the wish of the Liverpool members (both Founder Fellows and otherwise) that some clause should be inserted in the proposed new Constitution reasserting the status of Founder Fellows, such as actually naming them in the new Constitution.

The Liverpool members were unanimous in their approval of this suggestion, especially, Colin Askham, who had reaffirmed his interest in the Society.

Edwards hasten to assure me that the vast majority of the London members had a very sincere appreciation of the work that had been done, and that there was no probability of London passing a resolution in favour of removing a Liverpool member's status. He agreed with me, however, that too many Honorary Fellowships had been given out, and with the coming into force of a new Constitution Fellowships should be revised.

Edwards went on to say that the main point of divergence between Liverpool and London was that we in Liverpool had been fighting a dictatorship, to get a bureaucracy, whereas London wanted a real democracy, which would sound communistic to Liverpool's way of thinking. The number of independent drafts of the Constitution had served a useful purpose in preventing the omission of certain points that had perhaps figured in some drafts and not in others. One or two points had been omitted from all drafts, such as the appointment of proxies, when voting. However, it was his opinion that any final version of the Constitution, should be "vetted" by a lawyer in order to ensure that the wording was actually a legal implementation of the meaning intended.

By the first week of December 1936, the Londoners had hammered out a new draft Constitution, which Edwards described as "almost perfect", as it gave an exceptional degree of democratic control without in any way hampering the work of the officers of the Society. It was felt that this was an appropriate moment to call a Special General Meeting of the Society, and this was held on 6 December 1936 at my home address, 46 Mill Lane, Old Swan, Liverpool 13.

I gave my views to the meeting on the draft Constitution, which seemed more like a synopsis than the full Constitution, which I took it to be! Lack of binding phrases might admit of more democratic control, however. Askham, in particular, was concerned about the position of Founder Fellows in the future, and ways and means were discussed as to how best to safeguard their interests. If the London membership would not agree to Founder Fellows being named in the new Constitution, it was realised that there was little that could be done about it apart from giving Liverpool votes in favour of the proposal.

At the Special General Meeting a resolution was passed:

> "That the Society's Annual General Meeting for the year 1937 shall be held at a time and place to be determined by the London membership".

Askham was opposed to the idea of a transfer of the Society Headquarters to London, and was the only member present who voted against the motion.

He also wished to have the position of the Vice Presidency safeguarded, but the motion was put to the meeting and was duly carried by a majority against one - with my mother (who had been very worried about the amount of work I had been undertaking) chipping in to give her (unofficial!) vote for the motion as well.

In writing to Edwards, I stated that I would prefer to see a Vice President in Liverpool ("We might leave Askham in that post"), and in fact I couldn't see why a Society, couldn't have a dozen Vice Presidents, if necessary. There could in fact, be one Vice President to each branch of the Society, and this one could be nominated as President of the branch. But of course the actual number of Vice Presidents would depend upon the number nominated in the Constitution, and this, as well as nominees for the post would have to be decided upon at the Annual General Meeting.

So the matter rested, pending a decision upon the actual date of the Annual General Meeting. In the meantime, news from London was that Mr T Stanhope Sprigg editor-elect of Newnes' proposed science fiction magazine, had folded his tent and had stolen away into the night.

Seeker of Tomorrow had been returned with thanks, and (according to Ted Carnell) Maurice Hugi was "spitting blue lightning". He had had a number of stories accepted by Sprigg, and had that day completed 18,000 words out of a 20,000 worder that Sprigg had commissioned. However, three other publishers had been after Gillings to produce a British science fiction magazine, and he would be going into the matter with each of them. In the meantime, he was determined to bring out his fanzine "Scientifiction" as soon as possible. Eric Russell had some consolation for the return of the MS of *Seeker of Tomorrow*, inasmuch as his first story to be accepted, *The Saga of Pelican West* had been taken by *Astounding Stories*, and he had been paid $18 "on the nail" and before publication.

Summary of Events - The year - 1936

- The prospect of *Rockets through Space*
- Special forms printed - application form and details of membership
- Out of a membership of 80 - 20 Honorary Fellows
- Finances low - plus a Research Fund of £3-9-9
- J. Happian Edwards joins the Society
- J. Happian Edwards and P Cleator
- J. Happian Edwards' enthusiasm
- No London branch - Edwards given the address of Addey and Strong
- Arrangements with American and German Societies for exchange of publications
- The February 1936 issue of the *Journal*
- The Air Ministry "Rocket Ship"
- Willy Ley in the USA
- P Cleator becomes Honorary Member of the American Rocket Society
- *Rockets Through Space* was published towards the end of February 1936
- The value of *Rockets Through Space* to the Society
- The Annual General Meeting 1936
- A Vice Presidency offered to Professor A Low
- Professor A Low and Colin Askham
- The mystery of Ralph Stranger
- Hope of a monthly magazine to feature the BIS
- World Radio for 13 September 1935 and the BIS
- A separate journal for the World Wide Radio Research League
- A merger of three societies?
- WRRL publication to feature the BIS?
- Ralph Stranger hints at a change of name for the BIS
- Ralph Stranger's *Science Review* to appear in November 1935
- BIS participation in the *Science Review*
- The idea of the *Science Review* dropped

- Three issues of *Science Review* appear without the BIS
- Misunderstanding between Cleator and Stranger
- Question of payment for articles published in Ralph Stranger's *Science Review*
- Sketch of BIS badge sent to Ralph Stranger
- Russell and Johnson visit London, 25 March 1936
- Russell and Johnson to meet Carnell at Cannon Street Railway Station
- To meet at Liverpool Street Station instead
- Russell and Johnson wait for Carnell at the wrong station - Broad Street
- Carnell finally found at Liverpool Street Station
- Russell, Carnell and Johnson proceed to Ilford to see Walter Gillings
- Future if the meeting had not taken place
- Portents for the future
- The difficulties of finding Broadcasting House
- Russell and Johnson see Ralph Stranger and T Stanhope Sprigg
- Progressive forces at work in the BBC
- Cleator and Johnson contribute to Ralph Stranger's *Science Review* in *Ad Astra*
- Rushing to a climax with Ralph Stranger.
- Johnson to edit and produce the *Journal* of the BIS
- Free advertisement in the *Journal* offered to Ralph Stranger for *Science Review*
- Address of the Society now Mill Lane
- Ralph Stranger complaints about my secretarial notes
- *Ad Astra* - edited by P Cleator
- P Cleator fails to edit my rough notes
- Report on the Annual General Meeting 1936 omitted from *Science Review*
- Loan of block of BIS badge requested by Ralph Stranger
- ?Block a gift: ? Can reprint Stranger's articles "A Two Year Old Mystery" in the *Journal*
- Dr Karl Jansky's discovery

- Readers of World Radio asked to repeat Dr Jansky's experiments
- Letters from Batten - the block is a gift from Ralph Stranger
- Ralph Stranger to approve of draft article
- Loan requested from Stranger to astronomical block, to illustrate "A Three Year Old Mystery"
- *Requiem* for *"Ad Astra"*
- *Ad Astra* under a new name.
- *Ad Astra* No 3 sent to Stranger
- Ralph Stranger's *Science Review* suspended
- Ralph Stranger thrown by his horse
- The WRRL also suspended
- P Cleator resigned from the Council of the Society
- Reason for Cleator's resignation
- The Constitutional position
- Definition of the duties of a President
- P Cleator's secrecy
- Interests in the BIS is at heart by all
- Explanation to P Cleator
- Plea to Cleator for cooperation
- Cleator's reply and difficulties
- List of those to whom the *Journal* should be sent
- Proposed article in the *Journal* re Willy Ley's experiments in the USA
- Free advertisements in the *Journal* for Chambers' Dictionary and Chambers' Encyclopaedia
- Free advertisements in the *Journal* for *Rockets through Space*
- Difficulty of President who is not a member of the Council
- Cleator's comments re "Dictatorship and the Council".
- Cleator to be regarded as President until the next Annual General Meeting.
- P Cleator objects to being "regarded" as President
- Cleator - ? Life President
- Division of the spoils
- Eric Frank Russell is in limbo
- Science fiction and rocketry

- Walter H Gillings and rocketry
- 1000 envelopes for Gillings and "Scientifiction"
- Arthur Clarke's 10% of Russell's story
- Arthur Clarke's technical inquiries
- Arthur Clarke as a prophet
- Arthur Clarke 26th out of 1500.
- J Edwards and his advertisement in the *Journal*
- Letters from Edwards re London meeting
- J Edwards local representative
- Ralph Stranger - "well thought of"
- The June 1936 issue of the *Journal*
- Twenty four new members
- Congratulations from Arthur Clarke
- The only response received by Edwards - enquiries re health of Ralph Stranger
- Johnson misses Edwards at Euston Station
- Arthur Clarke moves to London
- Eric Russell to visit London
- Russell also to see Sprigg re *Seeker of Tomorrow*
- Arthur Clarke ambitions
- Walter Gillings - "science fiction first".
- I prepare to "Blow up"
- Voices in the science fiction "wilderness"
- Carnell is in state of euphoria following his meeting with Russell and Johnson
- Carnell visiting Gillings and Len Kippen
- Carnell thanks Johnson
- Carnell praises *Rockets Through Space*
- The use of magnetism in space.
- Carnell meets Edwards in spite of "Carnell's dance band"
- Carnell describes Edwards
- J Edward's "suicide club"
- J Edwards and atomic propulsion
- J Edwards and the sun
- J Edwards and time travel
- "The BIS could not be in better hands than Edwards'"

The Year 1936 | 145

- Johnson's aspiration for the BIS in London
- Johnson suggests gradual transfer to London, then the Annual General Meeting, 1937 to be held in the metropolis
- The Society's 100 members including 20 Honorary Fellows
- Russell meets the Londoners at Radiolympia
- Maurice Hanson/Dennis Jacques and *Novae Terrae*
- Questions re transfer of BIS HQ.
- Carnell and a matter of typewriting
- Next issue of the *Journal* delayed through lack of funds
- "Leave the *Journal* with Johnson for now!"
- To settle £3 debt before publishing the next issue of the *Journal* - then the transfer of the Society to take place.
- P Cleator resigns as President
- P Cleator's resignation was accepted by the Council
- Arrangements for the London meeting.
- "Joe Lyons" part in the BIS
- London meeting fixed for Tuesday 27 October 1936
- Success of meeting in London where 22 attended
- Professor A Low proposed as President
- Small London committee formed
- Suggestions for London activities
- Miss Huggett to do secretarial work.
- The London gathering, and the idea of a transfer of BIS HQ to London
- Misgivings re such a transfer
- "Hopeless" positions in the Liverpool
- List of BIS Council members 1936.
- The pessimism of Walter Gillings
- A summary of events continued.
- Walter Gillings joins the BIS as a member in November 1936
- Matters at low ebb in Liverpool
- Difficulty in obtaining 50% attendance at Council meetings.
- Johnson acting as secretary, treasurer, editor of the *Journal* and as general factotum
- Liverpool members experiencing an anticlimax and losing interest in the BIS

- Johnson and his fiancée (Miss Crossen) to visit London members on Sunday 15 November 1936
- Carnell euphoria dissolves: "I'm the Les Johnson of London"
- Edwards divides the London secretarial work between Miss Huggett and Mr K Chapman
- Edwards' scheme for additional contribution in order to provide a regular club room
- Johnson gives practical advice to Cornell regarding administrative details.
- W Gillings and R Smith added to the London committee of seven
- Over 30 persons were present in the London meeting
- Carnell appointed Director of Publicity and Gillings and J Strong on the Publicity Council
- The debt of the Society owed to E Carnell in the early days
- Johnson and Miss Crossen meet the London members at the Mason's Arms
- Miss Irene Cloke attends to keep Miss Crossen company
- Arthur Clarke sees Johnson and Miss Crossen back to Euston Station in the early hours of the morning
- Arthur C. Clarke appointed Hon Treasurer for the London branch of the Society
- "Proceedings of the London branch of the Society" covering the first three meetings of the branch
- A copy of the latest draft Constitution of the Society is sent to Mr J Edwards
- The Liverpool members ask that safeguards should be incorporated in the new Constitution reasserting the status of the Founder Fellows
- J Edwards: "No probability of London removing a Liverpool member's status"
- But: "The position of Hon Fellowships should be reviewed"
- Edward's views on the democratic procedure as related to the new Constitution
- Londoners' new draft Constitution "almost perfect"

- A Special General Meeting of the Society, called for 6 December 1936
- With only Askham in opposition the Special General Meeting voted that the next Annual General Meeting should be held in London
- Johnson stresses to Edwards that Askham should be retained as a Vice President
- Messrs Newnes drop the idea of a British science fiction magazine and the manuscript *Seeker of tomorrow* by Russell and Johnson is returned
- Gillings pushes on with his fan magazine scientifiction
- Eric Frank Russell first published the story is accepted by *Astounding Stories*

THE YEAR 1937

Preparation of the next issue of the *Journal* was in its final stages, and publication was held up only because I was awaiting the return of certain items I had left with Arthur Clarke while on my recent visit to London. By the end of 1936, the several Liverpool amendments that had been proposed to the new Constitution had been accepted at the London end and the Annual General Meeting had been fixed for 7 pm on Sunday 7 February 1937 at the Mason's Arms, Maddox Street.

As it happened, the hiatus, that would otherwise have existed between the Special General Meeting held on Sunday 6 December 1936, and the Annual General Meeting to be held on Sunday 7 February 1937, was to be time gainfully employed by at least those BIS members in Leeds, Leicester, Liverpool and London, who also professed an interest in the field of science fiction.

The first British Science Fiction Convention had been raised by D Mayer and the Leeds Chapter of the science fiction league, to be held in Leeds on Sunday 3 January 1937. Other chapters of the league existed at that time in Nuneaton (Leicester), Glasgow, Belfast and Barnsley. Enthusiasts from all parts of the United Kingdom and Northern Ireland were cordially invited to attend the Convention.

I was informed that Carnell and Gillings from London were unlikely to be able to attend because of the late hour at which the trains would return to London, and in addition, the cost of the trip, £1-3-3 (about £1-16) was somewhat prohibitive.

I was very fortunate myself from the point of view of convenience of travel, inasmuch as Eric Russell was good enough to drive me to Leeds in the car he had used for commercial travelling. It was as well that in the end, Carnell and Gillings as well as Clarke found it possible to be present otherwise, the only non-Leeds fan apart from Russell and myself to attend would have been Maurice Hansen from Nuneaton.

The Convention, which in many ways, could scarcely have been hailed as a great success, was nevertheless notable on two points:

1) It was the first British science fiction convention ever to be held

2) In spite of the fact that only six enthusiasts attended the Convention from outside the Leeds area, the Convention resulted in the formation of the pre-war Science Fiction Association.

With the formation of The Science Fiction Association in January, and the proposed transfer of the British Interplanetary Society from Liverpool to London in the following month, the pattern of events in interplanetary and science fiction from that time until the outbreak of war in September 1939 was beginning to become evident. The six "outsiders" at Leeds were of course, not only science fiction fans but were enthusiastic members of the BIS, and this was to illustrate how in pre-war days at least, interest in science fiction and interplanetary travel went hand in hand.

The London contingent returned home from Leeds determined not to embark upon such an expedition again (except perhaps to visit Liverpool!), then they sat back to await events at the Annual General Meeting. Meanwhile, Edwards was very concerned about my idea of mailing copies of the next issue of the *Journal* with the notices convening the Annual General Meeting. The official notices were in fact sent out to BIS members on 29 January 1937, together with a separate sheet, comprising the agenda, and Edwards was very relieved to be able to write to me in his letter dated 1 February 1937, to the effect that the notice had been received by him insufficient time.

The notice and the accompanying agenda read as follows:

---

29 January 1937

Dear Sir/Madam,

The Annual General Meeting of the Society for the current year will be held at the Mason's Arms, Maddox Street, Regent Street, London, W1 commencing at 7 pm on Sunday, 7 February 1937.

The agenda includes consideration of the proposed new Constitution of the Society (drawn up by a special committee of the London branch) and of a motion put forward by the Hon Gen Secretary on behalf of the Liverpool membership that the administration of the Society shall be taken over forthwith by the London membership.

Further proceedings, which include the presentation of the Financial Statement for the year 1936, will depend mainly upon the reception accorded to the proposed new Constitution of the Society.

The attention of members is drawn to the following bylaws in connection with the Annual General Meeting:

"No..matter..shall be...considered or discussed or any resolution is moved at an Annual General Meeting unless notice of the intention to bring forward such a matter for consideration shall have been given to the Secretary in writing five days at least prior to the date fixed for the holding of the meeting."

"None but members and officers of this Society or persons specially invited by the Council shall be present at any General Meeting of the Society".

"No member whose subscription is in arrear shall be entitled to be present and speak or vote at any General Meeting".

"Associate members have the same privileges accorded to members, accept the power to vote"

Yours faithfully

L Johnson

Hon General Secretary

PS, It is proposed that a deputation from Liverpool should travel to London for the Annual General Meeting. A train leaves Liverpool station at 10 am on the 7 February, fare 10s 6d return, returning from Euston Station, London at 12.35 am. on Monday and arriving at Lime Street at 4:50 am. All who wish to travel should inform me as soon as possible by P.C. when they will be informed of the arrangements that have been made.

LJJ

Annual General Meeting

7 February 1937

Agenda.

1. Notice convening the meeting

2. Appointment of Chairman

3. Minutes of the last Annual General Meeting

4. Report of Council for the year, 1936.

5. Report of the Hon Treasurer for the year, 1936

6. Consideration of the proposed new Constitution of the Society, prepared by the London Branch of the Society with the approval of the Council. (Letter from Professor A Low)

7. Appointment of officers and Council as directed by the Constitution, after consideration of item 6 on the agenda

8. Proposal by Mr L Johnson, Hon General Secretary of the Society, on behalf of the Liverpool membership: "That the headquarters and general administration of the Society be taken over by the membership of the London area"

9. General

L Johnson

Hon General Secretary

The British Interplanetary Society

Before the meeting, Edwards wrote to state that he would be interested to receive a list of whatever work I might be prepared to do for the Society in the future – this was for the benefit of the committee allocating appointments – and he urged me to try to get Cleator and Askham to attend the meeting, as he would like an opportunity to convince them that the Londoners were not really as bad as Cleator and Askham feared. Some of the Liverpool members including Colin Askham, had expressed their intention of attending the Annual General Meeting, but those who could not attend were eligible to appoint proxies or could propose written amendments to the Constitution.

As it turned out, the only Liverpool member to accompany me on the 10s 6d (52½) return trip to London for the meeting was Norman Weedall, who was, of course, Hon Librarian and a member of the Council of the Society. Strangely enough, in real life there seems to be only a story to be told if things go wrong. Nothing went wrong at the Annual General Meeting. All the necessary business was easily and smoothly transacted, and Norman and myself arrived back in Liverpool at about 5 am on Monday. After about three and a half-hour of sleep, I had to make my way by tramcar to my day's work at the Education Offices.

Unfortunately, the February 1937 issue of the *Journal* was not available either to be sent with the notice convening the meeting or even for distribution during the meeting itself. In fact, the issue was not sent out to members until the end of February. This was the second issue of the *Journal* of which I had the honour to have been editor, but the valedictory editorial was in fact written by Eric Frank Russell on my behalf, and I think it is worth quoting, as follows:

Editorial to the Journal of the British Interplanetary Society February 1937.

While expressing regret for the delay in the appearance of this issue of the *Journal* we feel justified in repeating the explanation already conveyed to our members in a special notice. This number has been held over pending completion of various arrangements that have had to be made in view of a probable transfer of the administration to London.

Throughout the past year, the Society has more or less paralleled the world's experiences. 1936 started on a note of considerable optimism, based upon the appearance of Mr Cleator's book, *Rockets Through Space*, and upon the hope of conducting practical experiments before the year had come to an end.

The formation of a new, and we believe more democratic Council, the publication of *Rockets Through Space* and the birth of a bigger and better *Journal* promised exceedingly well for the future. But, alas, we had wrongly drawn the shape of things to come.

Upon our overworked Hon Secretary descended a flood of new members attracted by the Council, the larger *Journal* and Mr Cleator's book. It is impossible to estimate the relative drawing power of these three factors, but in justice to our late President, we admit our suspicion that *Rockets Through Space* exerted the strongest pull.

After the sunshine the sunburn. Correspondence piled up to a height that one spare time worker could not reduce. The Society's greater strength reduced the ratio of the Merseyside membership to a mere one-sixth of the whole, and the tail was wagging the dog with a vengeance.

Our London branch, numerically twice as strong and undoubtedly much better placed to interest influential people, has become our logical heirs. Members should have no difficulty in perceiving the importance of the following motion that was adopted at the Special General Meeting held on 6 December 1936 at 46 Mill Lane, Liverpool 13:

"That the Society's Annual General Meeting for the year 1937 shall be held at a time and place to be determined by the London membership."

Our Metropolitan colleagues have arranged for an Annual General Meeting to be held at the Mason's Arms, Maddox Street, Regent Street, London, W1 commencing at 7 pm on Sunday 7 February. Two matters of outstanding importance to the Society will be discussed at this meeting: it is proposed that the Society's headquarters be transferred from Liverpool to London forthwith, and that the Society shall accept a Constitution drawn up and tabled by a special committee of the London branch.

It seems likely that this will be the last editorial from a Liverpool pen and is not out of place to acknowledge our indebtedness to that tiny group of individuals who bought the Society into being and carried it through its first difficult stages. We on Merseyside, have such confidence in our London branch that we have not the slightest doubt that the transfer, if it is affected, will be a great step towards a much larger and more active Society.

The February 1937 issue of the *Journal* included photographs of both P Cleator and myself, as well as Dr Otto Steinitz, Chairman of the E.V. Fortschrittliche Verkehrstechnik, the last named in order to illustrate an article by Miss D Farmer on "The 50th birthday of Dr Otto Steinitz".

Illustrations incorporated in the *Journal* showed the sketch plan of a rocket engine, a rocket motor of the Cleveland Rocket Society on the proving stand, with a further sketch purporting to be a diagrammatic section of Dr R Goddard's rockets, illustrating J Edwards' review of Dr Goddard's publication, *Liquid Propellant Rocket Development*.

Further items featured in the *Journal* were - Criticism of "Things to Come" by D Mayer, "Prize-Winning Paper on Rocket Design" (reprinted from The Scientific American), "Doubts on the Theory of Orifice Design" by J Edwards, a review of "Rockets Through Space" by Eric Russell and "The Proceedings of the London branch of the Society" already quoted earlier in this narrative.

Clarke thought the February 1937 issue of the *Journal* was better even than the previous issue, but wondered what Cleator thought of Eric Russell's review of *Rockets Through Space*. Indeed, when Cleator saw Russell's contribution in manuscript form, he said to me, "You're not going to publish that are you"? And truth to tell, in both issues of the *Journal* that I had edited there were various items, including some from Cleator himself, which I would not have chosen to have published, had I felt that I had had a completely free choice. One might even have said that some contributions (whatever merits they may have possessed in another context) were printed more for sake of peace, than for any relevance they might have had to the aims of the Society.

In the meantime, Clarke had managed to acquire an old Empire typewriter formerly belonging to Walter Gillings, and had broken all records in learning how to use it. The only previous practice he was able to claim had been on the headmaster's machine at school - when the head wasn't looking. Having tried Pitman's Manual of Typing, he found that this only slowed him down so he abandoned any idea of technique and carried on typing by instinct.

Unfortunately, his possession of this infernal machine only enabled him to pressurise me further, and amongst other things, to reveal to him in reply to his many questions that I held the sum of £4-14-3 for the Research Fund, plus £1-19-7 in Society funds. Cleator held £0-4-11½ and Clarke £0-11-5 making a total of £7-10-2½, held all in all on behalf of the Society.

The new Constitution of the Society was duly adopted at the Annual General Meeting held in London on Sunday 7 February 1937 read as follows:

Objects and Aims of the Society:

The British Interplanetary Society is a scientific organisation, whose activities embrace research in all problems pertaining to the conquest of space, and the realisation of man's age-old dream of interplanetary travel.

The immediate aims of the Society are the stimulation of public interest in the possibility of interplanetary travel, the dissemination of knowledge concerning the problems in which epoch-making achievements of extra-terrestrial voyages, and the conducting of practical research in connection with such problems.

The Journal of the Society:

All members receive free copies of the *Journal*, which is published as frequently as funds permit. It is ultimately hoped to produce a regular monthly issue.

The Bulletin of the Society:

Until such times as the *Journal* becomes a monthly publication, Bulletins are issued in order that members may be kept in touch with current events, both at home and abroad. The *Bulletin* takes the form of a duplicated report. It is issued free to members only.

Contemporaneous Organisations:

The Society maintains close contact with kindred organisations and their leading members throughout the world. Many renowned foreign experimenters have freely placed their services at the disposal of the Society. Noted foreign Fellows include Herr Willy Ley (Germany); Dr Jakow Perlman (USSR) Robert Esnault-Pelterie (France); Ing. Baron Guido Pirquet (Austria); Messrs. G Edward Pendray and Ernst Loebell (USA).

Constitution of the Society (7 February 1937)

1) The British Interplanetary Society shall be a scientific organisation to stimulate public interest in the possibility of interplanetary travel, promote and facilitate research into all problems associated with the conquest of space, and encourage, promote and direct all matters pertaining to extra-terrestrial development, and to carry out any acts it may consider contributory to the above.

2) The activities of the Society, shall be directed by a Council, which hall consists of the officers of the Society who shall be appointed by the members' committee (defined in clause 12). The Council shall coordinate the activities of the officers in the best interests of the Society as a whole, and shall delegate to such office as it considers suitable, any duties which may arise, and are not provided for specifically herein. The duties of the officers shall be as follows:

3) **President:** shall preside over the activities of the Society To be Chairman at any meeting at which he is present.

4) **Vice Presidents:** shall be elected to assist the President in his duties.

5) **Secretary:** shall receive and make communications on behalf of the Society, and shall prepare minutes of the meetings.

6) **Joint secretary:** to assist the secretary in the execution of his duties. To organise meetings and other gatherings of the Society.

7) **Treasurer:** to have charge of the monies of the Society, and to be responsible to the Society for them. To prepare an annual balance sheet and to be prepared to present the account when called upon by the Council. All monies except investments and a sum not exceeding £5-0-0, shall be kept at the bankers of the Society, in the name of the Society, and payment thereout shall be made by cheque on such bankers, signed by the Treasurer and countersigned by two other officers for discharge of such liabilities of the Society as shall severally exceed £5-0-0.

8) **Publicity Director:** to organise and control publicity on behalf of the Society. To keep a list of publications of interest to members of the Society. To be responsible for the publication of the *Journal* of the Society. The MSS of the *Journal* must be submitted to the Council for their approval.

9) **Research Director:** to organise research and investigation into matters of interest to the Society and render reports. To promote such development as is indicated by the results of the research, and to take such steps as may be necessary to secure the interests of the Society in such inventions, and the like as many eventuate.

10) **Librarian:** to be responsible for the safe custody of the literature of the Society. To collect such information as may be of use to the Society, from external and internal sources to acknowledge in the *Journal* the receipt of matter, which may be donated to the Society.

11) **Members' representative:** to receive suggestions from members and to see that they are bought to the notice of the officer concerned and to report back to the member concerned in what manner the said officer has dealt with the matter.

12) **The Members' Committee:** hereinafter referred to as the Committee, shall consist of seven members who are elected by the members present at every monthly general meeting. No member may sit on the committee for more than 12 months out of two years.

13) **The Committee** may recommend the replacement of any Council member, subject to ratification by the vote of the next general meeting and subject to due notice having been given to the membership.

14) **The Council** can override the actions of officers in their duties, by an otherwise unanimous vote. No action of the Council shall be effective unless due notice has been given beforehand to the members' Representative.

15) **Active membership:** shall be open to any person acceptable to the committee. The subscription for membership shall be 10s 6d annually or 3s 0d quarterly.

Membership entitles the holder of the membership to be present at, and to vote at, all general meetings, and to receive all publications of the Society. Membership carries with it the duty of accepting any office which may be offered or assisting any officer when called upon, to the best of the member's ability, except some sufficient reason to the contrary be accepted by the committee. The right to membership may be rescinded at any time by the joint unanimous vote of the Council and the Committee.

16) **Associate (or Corresponding) membership:** shall carry all the privileges of membership, except the vote, and shall carry no duties. The subscription for associate membership shall be 7s 6d annually or 2s 0d quarterly.

17) **Associates** shall be persons who are tentatively interested in the activities of the Society, and shall receive the *Journal* for a fee of 2s 6d per annum.

18) **Arrears.** Any person more than six months in arrears with subscriptions shall be deemed to have left the Society, notwithstanding, that this may not be deemed in any way rescinding any just dues that may be allowed to the Society by the members.

19) **Fellowship** shall be an honour bestowable by the committee and shall be purely an honour, carrying no special privileges except a free subscription to the *Journal* for such period as shall be decided by the committee.

20) The following are **Founder Fellows**, and shall, after the President and the Vice President, have precedence to preside over meetings at which they may be present, provided that they shall continue to be active members of the Society:

C Askham (G6TT), H Binns, P Cleator AMIRE, AMIET, FRSA,

J. Davies (G20A), W. Dunbar, J. Free Jr., L. Johnson, T. McNab and N. Weedall.

21) **Branches:** Any seven active members in one place, may, subject to the agreement of the Council, form a branch and elect a Branch Chairman, who shall control the branch inside the limits of the Constitution and the Society.

22) **Meetings**: An Annual General Meeting shall be held at such time and place as the Council shall determine, when the officers shall report on the year's activities. Notice shall be given in the *Journal* and all members shall be duly notified. A monthly general meeting shall be held at such time and place as shall be decided upon by the Joint Secretary.

23) **Quorums shall be as follows:**

Quorum of AGM 13

Quorum for Amendment of Constitution 13

Quorum of Council 7

Quorum of Committee 4

Quorum for the election of Committee 13

In the event of a quorum not being obtained there shall be no alteration in the status quo. There shall be no Committee election unless it is intimated to the Chairman of the meeting that a member or members is, or are retiring or challenged when the Chairman shall ask for nominations, which must be given immediately, and a vote taken as to which of the nominees shall fill each vacant or challenged seat.

24) **Proxy votes.** Anyone member may represent by vote not more than two absent members. A signed authorisation from the members who are being represented must be produced by the member representing them. Proxies do not count in establishing a quorum.

25) **Winding up.** In the event of a Society being wound up, any assets in its possession shall be divided equally among the active membership. No member shall be liable for more than an amount equal to his annual subscription.

26) **Amending the constitution.** Any amendment to the Constitution must be proposed by an officer or committee member, and require a two to one majority to pass it, at a general meeting specially convened for the purpose after due notice.

Reading the Constitution now, it seems to be a little odd that at the Annual General Meeting, when under most constitutions, the Council and Officers of the organisation would stand for re-election, under this particular Constitution, the officers were merely to report upon the year's activities. Furthermore, the Officers and Council of the Society would appear to have been appointed by and to have been under the authority of the members' committee the actions of the Council being subject to due notice having been given to the members' representative.

In accordance with the provisions of the new Constitution, therefore, the members' committee met at the home of Mr J Edwards to elect the Officers of the BIS. Professor A Low, DSc was elected President, although, Mr A Janser had suggested Sir James Jeans - a noted popular astronomer of the time.

By unanimous vote, Mr L Johnson was elected a Vice President, "In view of his service to the Society in its perilous youth," According to Arthur Clarke. Walter Gillings was suggested for Publicity Director and Ted Carnell Joint Secretary plus Social Organiser. Miss Huggett, Edwards and Clarke retained the appointments they had held under the auspices of the London branch (as it had existed previously), Mr A Janser volunteered to become Hon Librarian.

It will be noted that the new Constitution abolished the idea of being able to buy a Fellowship for £2-2-0 per annum, and Fellowship in the future was to be purely an honour, carrying no special privileges except for a free annual subscription to the *Journal*.

In preliminary discussions relating to the possible transfer of the Society headquarters to London, I stated that I saw no reason why there should not be a new grade adopted for "Founder Fellow (London)", but this idea was not taken up by the London members.

Indeed, according to Arthur Clarke in his capacity as Hon Treasurer of the BIS, most of the original "Founder Fellows" could be deemed to have left the Society under the arrears clause of the new Constitution! I suggested that Fellows of all three varieties (Founder, Honorary and Paying) should have letters sent to them, explaining the new circumstances, and those who did not reply reiterating their interest in the Society should be crossed off the membership list.

In fact, I went on to say to Arthur that I thought that all members in arrears should be tracked down and urged to pay up, as 1,000 copies of the *Journal* were expected to arrive for distribution at any moment. Funds were low, and the production of the *Journal* had to be paid for.

I then learned that the mysterious H. Grindell Matthews had contacted the technical committee of the Society; Arthur Clarke thought Matthews was "most interesting" and a "live wire", possessing, as he was alleged to do, his £2,000,000 or £5,000,000 backing and his £3,000 laboratories. On the other hand, Clarke also thought that Matthews was a "merchant of death", with his chief interest being in weapons of war, and was "cute" enough to see the possibilities of the rocket - as the Germans had done four years earlier.

And so it came to pass, that I was indeed (as predicted) writing to Arthur Clarke (in his capacity as Hon Treasurer of the Society) instead of him writing to me.

There was some concern on the part of the committee, with regard to the position of P Cleator. It was felt that an honour of some kind should be bestowed upon him in order to show appreciation of the work he had undertaken as Founder of the Society. There was a suggestion that he might be prepared to accept a Vice Presidency, or it was thought that he might prefer to remain as a Founder Fellow without being involved in any particular office in relation to the Society.

It was thought by some members of this committee that having Cleator as a Vice President as well as myself might make my own position uncomfortable, and they were therefore seeking my views on the subject. I replied that some kind of an honour was certainly indicated for Cleator, but what could be done depending on so much on his own views, in what were admittedly difficult circumstances. I mentioned the possibility of him accepting a position as Vice President, or a number of alternative appointments including "Astronautical Advisor".

It was Colin Askham, however, that I was more concerned about than Cleator. I was in effect taking Colin's place as Vice President, and I did not see the need for this. Unfortunately, he was a comparatively unknown quantity to the London members who did not appear to realise the important role he had played in the continuation of the Society at critical stages in its development. All I could get out of my friend, Arthur Clarke, on this point was, "We'll see...".

As it turned out, Ted Carnell did not wish to become Joint Secretary, but accepted the post of Publicity Director, which he had previously held under the auspices of the London branch, with Walter Gillings as Joint Publicity Director. Mr R Smith accepted the position of Joint Secretary with Miss Huggett.

I was very pleased to note that on 17 February 1937, Phil Cleator in fact wrote to J Edwards to convey his thanks to the Members Committee for their offer of a post as Vice President, which he accepted "with pleasure".

After my frenzy to have the administration of the Society transferred to the London members, it may seem strange that after the resolution to do so had been so smoothly passed at the Annual General Meeting, the Londoners (especially Miss Huggett), seemed to be becoming a little restless over my delay in actually transferring the Society to London. Arthur Clarke, in particular, waxed humorously sarcastic over the delays.

However, it must be remembered that I had a full time job to hold down at the Liverpool Education Offices, working from 9-5 Monday to Friday, and even until 1 pm on Saturdays. There was also still much to be done. The 1,000 copies of the *Journal* had to be dealt with, involving envelopes to be addressed, filled, stamped and posted. The balance sheet and statement of accounts for the year 1936 had to be completed and audited.

Clarke humorously described my income and expenditure account as being merely a receipts and payment account. The question of a balance sheet had to be waived, with everything being in the melting pot due to the process of transfer. Before the accounts could be completed, the sum of £9-10-6 (£9.52½) had to be paid to the printers for producing the *Journal*, £0-12-3 (61p) to Messrs Roneo for packing and despatching the duplicator to Ted Carnell, then there was the cost of sending two large parcels by rail to Miss Huggett. A three-page letter also went to Miss Huggett explaining various points about the consignments.

By the last week of April 1937, I realised that the end had come at last and that everything that finally been cleared, including a further parcel of goods sent off to Miss Huggett. Several hundred copies of the *Journal* had been sent to editors all over the world. and books and cash was sent to Arthur Clarke.

Things had been shaping up well in London, despite their lack of administrative facilities; J Edwards had given a lecture on "How a Rocket Works" and Arthur Clarke had spoken on (or, as he put it, about) the Moon. It was with a feeling of some sentimentality that I finally found myself cut adrift from the management of the BIS, after nearly four years.

On receiving the duplicator and supplies, Ted Carnell had a rare tussle trying to produce the first issue of the *Bulletin*. Did he need more ink? Or should he have typed the stencil harder? Above all, he had considerable trouble with an automatic feed; he could not entice the machine to pick up each sheet singly and it kept lapping them up in bunches. Finally, he had to feed the sheets of paper through the machine one by one. It took a long time.

A visit by Ted to Roneo's revealed the fact that nobody there knew how to work a Model number 20. (it was in fact a number 10!) and I finally had to resort to typing out detailed instructions (as far as I could remember)- on how to do it. Having struggled valiantly, through the first few issues of the London version of the *Bulletin*, Carnell surprised all by suddenly producing a printed issue.

Whether this was triumph or failure on his part was a moot point. He was, of course, employed in the printing department of Gamages Store, so his encounter with the duplicator must have been particularly irksome. In any event, the printed version of the *Bulletin* (a single foolscap sheet) looked infinitely better than the duplicated version.

I pointed out in as kindly manner as I could that this *Bulletin (New Series)* was in fact the third in a series of *Bulletins* that had been issued by the Society, and that unfortunately the two large additions that had been issued by James A Free Jr. and myself at the end of 1934 had also been dubbed *Bulletin "New Series"*. Perhaps Carnell's should have been called *"Third Series"* or even *"London Series"*.

The transfer of the headquarters administration of the Society now having been completed, I can do a little better than conclude this major portion of my narrative by quoting from Ted Carnell's *Bulletin (New Series)* dated 4 April 1937, Volume 1, No 1 describing the transfer of the Society from Liverpool to London, as follows:

Important Notice

The outcome of the Annual General Meeting held in London on 7 February was, unfortunately, too late for inclusion in the last Journal. Members are particularly requested to note that on a universal agreement vote, the headquarters of the Society was officially transferred from Liverpool to London, and all business pertaining to the Society should now be directed to the new Officers at the address given below.

This does not mean that Liverpool members, and particularly L Johnson have left the Society. On the contrary, they will continue to operate as enthusiastically as before, but without the arduous task of managing the many positions which have arisen. It was found that in London there were many more members than in Liverpool with whom to share the increased work of the Society, and, being more centrally situated for contact with various people interested in astronautics, the obvious outcome was to transfer the headquarters to the South.

Officers for 1937

The result of the elections held at the Annual General Meeting is as follows:- President: Mr A Low, Vice Presidents: P Cleator and L Johnson, Hon Gen Secretary: Miss Huggett, Treasurer: A Clarke, Organising Secretary: R Smith, Research Director: J

Edwards, Publicity Director: E Carnell, Librarian: A Janser, Member's Representative: CG Smith.

The above Officers also comprise the new Council.

Bulletin and Journal

The *Bulletin* is being reissued with the hope of keeping in closer touch with members who live outside Greater London and are not able to attend the regular monthly meetings. The *Journal* will still be published at frequent intervals, when material warrants. Copy for both publications should now be sent to E Carnell, 17 Burwash Road, Plumstead, London, SE 18.

Hon General Secretary

The new address of the Secretary is now 92 Lockwood Road, South Chingford, London, E4. Miss Huggett states that the stamp account is now so high that it seriously handicaps the activities of the Society. Would members making communication require an individual reply please enclose a stamped addressed envelope.

Treasurer

Contributions should now be sent directly to the Treasurer in London, A Clarke, 21, Norfolk Square, Paddington, W1 when the Society's official receipt will be forwarded.

Research Director

J Edwards, with an address at 92 Lockwood Road, E4 hopes to keep members, unable to attend meetings, informed of technical progress through the medium of the *Bulletin*. To assist in the development of this service would all members submit to him the following information in the fullest detail. Any scientific subject of which they have knowledge, interest and/or facilities, and any questions upon interplanetary matters which they have been asked or upon which they themselves would be interested in having any information.

Librarian

A Janser of 28, Great Ormond Street, London, WC1 will be compiling the new library of the Society, and would be extremely grateful to any members who sent him scientific books which will be suitable, and which they have no further need for. These will be duly acknowledged and placed at the Society's disposal. It has also been suggested that a library of science fiction magazines should be compiled, details of which will appear later.

Members Representative

A new position was created to deal with queries between members and the Council, thus lessening much of the Secretary's correspondence. Members who have any suggestions (or complaints) which they would like brought to the attention of any officer should direct their query to C Smith, "Greengates", Manor Drive, Hinchley Wood, near Esher, Surrey, and Mr Smith will attend to the matter and report back to the member immediately.

The next monthly meeting

Mr R Smith the Organising Secretary, also using the joint address at 92 Lockwood Road, E4, notifies members that the next monthly meeting will take place on Tuesday, 6 April at Mason's Arms, Maddox Street, London, W1. The main subject of the meeting will be addressed by J Edwards "How The Rocket Works", which will last from 8 pm until 9 pm and a general discussion afterwards, until 10 pm. Committee and Council members are requested to attend at 7 pm enabling any formal business that they may have to be disposed of before the main meeting.

Technical Committee

A Technical Committee has been formed in London under the management of Edwards. They hope, at a later date, to issue a textbook of astronautics, which can be used as a reference in future experiments and will also be a help to members who are not so well informed on technical matters.

Get a friend to join the BIS, and support British Astronautics.

Summary of events: The year - 1937

- The *Journal* for February 1937 in the course of preparation
- The new draft Constitution is ready
- The Annual General Meeting is fixed to take place in London on Sunday 7 February 1937
- The first British Science Fiction Convention is arranged to take place in Leeds on Sunday 3 January 1937
- Only six non-Leeds fans attend the Convention
- The six Leeds "outsiders" were both BIS and SF fans
- Leeds - "Never again!"
- Notice re the Annual General Meeting received in good time by J Edwards
- Edwards would like Cleator and Askham to attend the Annual General Meeting
- In fact, Johnson and Weedall, the only Liverpool members to attend the Annual General Meeting
- All went smoothly at the Annual General Meeting
- Contents of the February 1937 issue of the *Journal* including a photograph of Cleator and Johnson
- Cleator re Russell's review of *Rockets Through Space*
- Clarke's typewriting
- £7-10-2½ held on behalf of the Society
- The new Constitution quoted in full
- Relationship between the members' representative, the members' committee and the Officers of the Society
- Professor A Low is appointed as President
- Other appointments include Johnson as Vice President
- The position of three varieties of "Fellow"
- Clarke re H Grindell Matthews
- Cleator's position discussed
- The position of Askham in the Society
- Cleator accepts a position as Vice President
- Delays involved in the business of transferring the Society to London
- Bills to be met before the transfer can be completed

- The duplicator is sent to London per Messrs Roneo
- The BIS in London takes shape
- Carnell tries to use the duplicator
- Carnell reverts to type
- The *Bulletin* - another *"New Series"* is issued by Carnell from London- giving details of the Annual General Meeting and the transfer of the BIS HQ in London
- The transfer of the BIS from Liverpool to London is finally completed.

AFTERMATH 1937 TO 1939

By April 1937, the transfer of the Society to the London members had been virtually completed. Unfortunately, after the first nine issues had appeared of the *Bulletin* it reverted to a single Quarto duplicated sheet. According to the first of the revised duplicated issues (Vol 2 No. 1 January 1938) "the printer interested in the Society could no longer manage to do the issue so cheaply." I have a sneaking suspicion, however, that Gamages printing department may have thought that Ted was doing far too much private printing in his lunch hour.

Ted Carnell had edited and produced the first 11 issues of the *Bulletin*, and may in fact have produced two more, but on these two issues (Vol 2 Nos. 3 - 21 March 1938 and 4 - 21 April 1938) the name of the editor is not given.

Vol 2, No. 5 -21 May 1938 - states "A Clarke, Treasurer", and Vol 2, No. 6, duplicated in black ink instead of the usual blue (and apparently not from the BIS duplicator) gives M Hanson as editor.

With Vol 2, No. 7 - August 1938, the scene changes: A Clarke is given as the editor, with an address at 88 Gray's Inn Road, London WC1. The following is quoted from his editorial:

> *Previous Bulletins* have been single sheets containing little more than a synopsis of the last meeting and the date of the next. Even so, their production was a considerable strain on the single officer responsible for them, who had not only to collect the material but also to duplicate the sheets and address the envelopes. Recently, however, three officers moved from outlying districts to more central quarters in London, and now find themselves in a position to produce a considerably larger *Bulletin*, running, we hope to eight or ten pages eventually. This *Bulletin* will contain not only full reports of all meetings, but also many other matters of current interest which would not be suitable for the *Journal*.

To members of the Science Fiction Association who were regular recipients of *Nova Terrae* the new format of the *Bulletin* (single folded foolscap sheets, duplicated in light blue ink) would have been very reminiscent of Maurice Hanson's science fiction magazine.

Effortlessly settling into his job as Hon Treasurer, Arthur Clarke hastened to remind me that a large number of Liverpool members were well behind with their subscriptions, and seeing that I was grieving at no longer being involved in the BIS administration, he gave me the somewhat dubious honour of chasing up the delinquents.

Roberts, Toolan, Davies, Free, Binns, Askham, Miss Hastie, Eddon, McNab, Hodgson, Beardwood, Weedall and Dunbar were all behind with their subscriptions to a greater or less degree and I told Arthur that in the past I had tried all I knew to persuade them to pay up, so far without success.

My advice to Arthur was to cross the first five names off the list of paid-up members; Weedall and Dunbar would probably pay up in due course, and Askham had always done so in good time. Eddon and Beardwood worked in the School Medical Department where I, myself, was employed, and had every opportunity to pay if they wanted to or were able to do so.

McNab seemed to have disappeared and may have gone abroad. Hodgson had said "No meetings, no subscriptions", being (I supposed) another version of "No taxation without representation" but he had been invited to attend the meeting that I had called for on 16 April 1937 and had not attended and had not sent any apology for not having done so.

To the names of the three Londoners (Carnell, Clarke and Gillings) who had attended the Leeds Science Fiction Convention, must be added the name of William Temple, as one who managed to reconcile the twin worlds of science fiction and interplanetary travel. Bill Temple joined the BIS on 26 August 1936 and had been present at both the meetings of the Londoners that I went down to address. Born in 1914 (the same year as myself), he had read most of the science fiction magazines since he had picked up a copy of *Amazing Stories* in 1927. He was a somewhat reluctant employee of the Stock Exchange and confessed that he was not qualified in any particular branch of science, but was anxious to do any general work that he could on behalf of the Society.

W. F. Temple had contributed an exhilarating series of articles to *Novae Terrae*, commencing with the April 1938 issue entitled "The British Fan in his Natural Haunt", his first subject having been Eric Williams, at one time, a Council member with myself of the Science Fiction Association.

Following the second British Science Fiction convention held in London on 10 April 1938 (and much better attended than the Leeds Convention - might we say!) it was typical of the close association between science fiction and interplanetary travel, that the man elected as president of the SFA was none other than the ubiquitous and genial professor A Low. And it was not remarkable that at the Annual General Meeting of the SFA held at the Convention that the members agreed to transfer the headquarters of the SFA from Leeds to London, following in the footsteps of BIS.

Others to be honoured by Bill Temple's attention in the series of articles were Ted Carnell, Arthur Clarke, Maurice Hanson, Ken Chapman and Walter Gillings. In a supplement to the last issue of the *Novae Terrae* dated January 1939 (after which Ted Carnell took over the Fanzine as "New Worlds") Arthur Clarke got his own back by adding a seventh name to the series by featuring Bill Temple himself in his natural haunt.

And he was well able to indulge in verbal caricatures of Temple, because in June 1938, Arthur had decided to move in with Bill Temple and share the flat at 88 Gray's Inn Road, WC1. In September 1938, Maurice Hanson who had previously shared a bed-sitter in Bernard Street with a Leicester school friend (having wangled a Civil Service appointment in the Metropolis) also moved in with Clarke and Temple, who by that time had thrown open the flat to SFA members on Thursday evenings.

It was not particularly surprising, therefore, to find that 88 Gray's Road became the combined headquarters of both science fiction and interplanetary travel, and from which was issued not only *Novae Terrae*, but also the *Bulletin* of the BIS. Indeed, Bill Temple was to go on to become Publicity Director of the Society in succession to Ted Carnell (who had also become Hon Treasurer of the SFA) and to succeed Ted as editor of the *Journal*.

Ted Carnell had resigned as Publicity Director in order to become the editor of *New Worlds*, and which was to feature science fiction stories rather than the science fiction fan activities that had been the highlight of Maurice Hanson's *Novae Terrae*; Ted had also become involved, with myself, in our firm of Science Fiction Service established for the sale of science fiction books and magazines throughout the world, by mail order.

It may have been a coincidence (but more likely part of The Plot) that the first British Science Fiction Convention in Leeds and the foundation of the Science Fiction Association took place in January 1937, while the transfer of the BIS HQ from Liverpool(following the formation of the London branch of the Society in October 1936) took place in February 1937; following the formation of the London Branch of the SFA in October 1937, it took only until April 1938 for the headquarters of the SFA to follow the headquarters of BIS to London.

The transfer of BIS HQ having taken place, I arranged for a meeting of the BIS/SFA to take place in Liverpool on 16 April 1937, but despite having sent out invitations to 30 individuals, only six turned up, including Eric Russell and myself. One fellow named O'Brien, "who knew little or nothing about anything", turned up to tell us he wasn't really interested. An appeal was made in *Novae Terra* for any fans in the Liverpool area to make contact with me with a view to enlivening (and repopulating) future meetings in the City.

Vol. 1, No 6 of the *Bulletin* of the BIS dated 26 August 1937, announced that the BIS was shortly to organise a competition for model aeroplanes using jet propulsion instead of propellers. Professor Low kindly offered a cup to be presented to the winner; those wishing to take their tests in the Liverpool area were advised to make arrangement arrangements with P Cleator at his address in Wallasey.

The possibilities of the model aeroplane competition prompted me once again to summon the local fans to a meeting, and this took place on 19 August 1937; over 18 persons attended, both BIS and science fiction enthusiasts, including P Cleator, who was accompanied by Miss Madeline Birmingham (possibly spelt "Bermingham") and later to become Mrs P Cleator. By "coincidence", Miss Birmingham was already known by my own fiancée, Hilda Crossen, both having been employees of the then embryo Vernons Football Pools at Russell Buildings, School Lane, Liverpool, in the early 1930s.

In the presence of the vast concourse (which included Walter Gillings on holiday with the Russells), I asked Phil Cleator, "Have you read our story"? referring to *Seeker of Tomorrow* by Eric Russell and myself that had been the feature story in the July 1937 issue of *Astounding Stories*. "Story?" asked Cleator, "when did it appear?" "In *Astounding Stories* for July", I informed him. Dawning comprehension in Cleator's eyes - "Oh, I never read *those* magazines!"

Wally Gillings' visit was particularly opportune, as he had just realised his great ambition to edit a British Science Fiction magazine; Messrs

World's Work (1919) Ltd had agreed to a trial issue of such a magazine, being produced, to be called *Tales of Wonder and Super-Science*. Following the relative success of the first issue, 16 issues in all were to appear dated between Summer 37 and Winter 1942.

Relieved of BIS work I was able to find time to indulge in some writing again and had a short article, the "Transatlantic Rocket Mails" accepted and printed by the September 1937 issue of *The Meccano Magazine*.

Following this small degree of success, I had a short story, "Satellites of Death" published by Water Gillings in *Tales of Wonder* dated Summer 1938. However, seldom loath to find work for my never idle hands, I then devoted most of my time to "The Science Fiction Service".

From the end of the year, 1935, I had organised a small mail-order Science Fiction book and magazine business, operating in the name of my brother, Victor Horatio Johnson. When Ted Carnell visited me in August 1937, it was agreed that he and I should form a partnership in this enterprise with branches in Liverpool and London.

Thus, we started the Science Fiction Service. In September 1938, we actually opened an office in Liverpool in Room 7 in No 15 Houghton Street. This cost us a rental of £0-7-6 per week, and was situated centrally in the city, opposite the side entrance of Messrs Owen Owen's (one of the large departmental stores) and above Foners Corset shop (now in Bold St). The exotic displays in the window of Messrs Foner's used to regale our visitors from the far corners of the kingdom.

In June 1938 a Liverpool Branch was formed of the Science Fiction Association. Early meetings were held at the old BIS rendezvous, the Hamilton Cafe, then at a famous Liverpool lunchtime eating house, Messrs G Petty's Cafe in Hackins Hey. Then, just as "the flat" had become the centre for BIS/SF fans in London, so the Office became the focal point until the outbreak of War, for fans from Liverpool and elsewhere.

It was at the Office that we produced the early issues of our SFA sponsored *Science Fantasy Review*, of which I edited the first two issues, and which was continued into the war years as *Science Fantasy Review War Digest* by Ronald Holmes.

Meetings of the BIS/SF fans were held at frequent intervals, including new enthusiasts, who were to make their names, mainly in the field of SF. Such were Leslie Heald (Charnock Walsby), R Holmes himself, (*The History of Witchcraft in Britain*), John Burke (*Chitty Chitty Bang Bang*), David McILwain (Charles Eric Maine), H Dickinson (*The Sex Serum*, etc. and of Red Letter fame - if that is the word!) and Ernest L Gabrielson (sponsor of *The Viador Fellowship*) - not to forget Frederick Bowman, a legendary Liverpool figure.

The appearance of *Tales of Wonder* had apparently encouraged Messrs Newnes to revise their attitude to Science Fiction and in July 1938 they produced the first issue of *Fantasy*, edited by T Stonehope Sprigg; however, this was only to see three issues before being discontinued in July 1939. We even had a kind of local Liverpool triumph, when the third issue of *Tales of Wonder* appeared, featuring not only my own *Satellites of Death*, but also *The World's Eighth Wonder*, by Eric Frank Russell, and *The Giant Bacillus* by H Dickinson.

No 15 Houghton Street has long since been demolished, but it is perhaps significant that the former site of the Office is now clearly visible for miles around Liverpool, and even out to sea. For rising right through the centre of what had been No 15 is The Tower Restaurant, thrusting its umbrella shape 450 feet into the air above the remains of Houghton Street, and looking for all the world like a giant HG Wells type spaceship, about to blast off from the Earth.

It was nearly a year after this transfer of the BIS to London, that the first London edition of the *Journal* appeared dated December 1937, and was edited by the hand of my good friend, Edward John Carnell. Two issues per year, in addition to the monthly *Bulletin* had been the target set for themselves by the Londoners, but despite the ever-increasing membership (still a long way short of the 200 mark by August 1939), they found themselves beset by the same financial problems as the Society in Liverpool. Thus, their second issue of the *Journal*, edited by William F. Temple dated December 1938, did not in fact appear until January 1939. Indeed, by the time of the outbreak of the war in September 1939, not more than three issues of the *Journal* had been published, with dates in 1937, 1938 and 1939.

However, a much greater emphasis had been placed by the London group on the experimentation and design, as far as possible within the limits of their resources. Apart from their competition for a model jet-propelled aeroplane and a proving stand, they had in hand the design of a coelostat, an altimeter, a primary battery, and (actually) a BIS spaceship, which was to consist of a detailed feasibility study of a three-man lunar vehicle.

The London version of the *Bulletin* had followed very much in the footsteps of its Liverpool predecessors, giving details and meetings and lectures and debates and exhortations to members to pay their subscriptions. In addition, it included articles on various aspects of interplanetary travel, such as "Limitations of Space Travel" and "The Temperature of Space" by Friedrich Schmiedl, "Rocketry and the Law" by Eric Burgess and "The Planet Mars" by L Carter.

Unfortunately, Arthur Clarke's efforts with the *Bulletin* were not appreciated by everybody. Cleator's view (expressed in the columns of *Tomorrow*) was that mimeographed reports were not worth one-tenth of the time and trouble that their preparation involved. And Eric Russell, in his turn, condemned the January 1939 issue of the *Bulletin* as "unusually lousy". (*Tomorrow* was D Mayer's high class printed science fiction fan magazine, published in Leeds, and with which Walter Gillings' "Scientifiction" had become incorporated when Gillings became the editor of *Tales of Wonder*).

In his editorial in the August 1939 issue of the *Bulletin*, Arthur Clarke had explained that the *Bulletin* had been issued earlier than intended in order to announce the next meeting of the Midlands Branch of the Society, and also to obviate production difficulties when the editor was away in September.

He announced that the following issue of the *Bulletin* would probably be out during the third week of October, "though as we are now in the midst of the annual (war) crisis" it was impossible to be definite on this or any other point. He only hoped that the members did not receive the August 1939 issue of the *Bulletin* – if indeed they received it at all – by hand messenger sometime in 1940.

In the meantime, Hilda Crossen and myself had duly been pronounced man and wife during a ceremony at the church of Christ the King, Queens Drive, Liverpool on Saturday 26 August 1939, then departed for our honeymoon in the direction of Llandudno, North Wales.

Alas for the plans of mice and men! Within five days, Hitler's armies had invaded Poland (during the beautiful summer weather then prevailing), a War Emergency and Blackout had been imposed throughout Britain and Hilda and myself scrambled back on the Friday after our wedding – in the dark – to a sparsely furnished house at 16 Wilton Road, Huyton, Lancashire – a few miles outside the Liverpool City Boundary.

At 11 am on Sunday 3 September 1939, with our marriage only just over one week old, Britain in the shape of Sir Neville Chamberlain declared war upon Germany. I reported back to the Education Offices, voluntarily, in order to assist with the arrangements for the evacuation of schoolchildren.

The August 1939 issue of the *Bulletin* was the last to be issued for over six years; in October 1939, a foolscap size, single sheet circular (apparently from the hand of Arthur Clarke, and in the familiar blue ink of the famous BIS duplicator) was mailed to all members of the Society. It read as follows:

Emergency Meeting

An Extraordinary General Meeting of The British Interplanetary Society took place at 88 Gray's Inn Road on Tuesday 10 October. The meeting was held to consider the best means of carrying on with the Society's work.

It will be realised that the dislocations caused by the war will render it impossible for the BIS to continue with its main activities. The majority of our officers are liable for military service, and difficulties of communication caused by transfer and evacuations have already made it impossible for the Council to continue meeting. These difficulties will increase rather than diminish, and consequently, we have no alternative but to suspend activities for the duration of the war.

We have, however, no intention of abandoning our work, especially since so much has already been accomplished. Although for the present all official activities must cease, plans have been made for the safe custody of the Society's records until such time as it will be possible to continue once more with our work. On the termination of the war, we will endeavour to get in touch with our members by writing to their present addresses. If through a change of address, any member does not hear from us, he should write to:

The British Interplanetary Society, "Ballifants", Bishops Lydeard, Taunton, Somerset. Letters to this address will, we hope, be collected eventually by some members of the Council. Write to this address only if you have not heard from us within a reasonable time after the end of the war, and do **not** address your letter to any officer personally. Alternatively, we would be glad to receive from members, new addresses, other than those they have already given us, to which it might be better if we sent our first post-war notice.

At the moment, the Society's financial assets consist almost entirely of the accumulated balance of the Research Fund, and it is comparatively little in respect of unexpired subscriptions. Accordingly, we intend to place this balance on deposit at our bankers, so that, at the increased bank rate, it should earn quite a reasonable amount of interest. During the war, and the probable absence of the Society's executive, expenditure will drop to a very low level and it should be possible to re-date members' subscriptions so that they will receive the full 12 months of service in respect of every year for which they have paid. Thus, if you paid a year's subscription three months before the outbreak of war, you will be credited with about nine months from the date when the Society commences its activities again.

These arrangements, which have the full approval of the President, the available officers and the members who attended the last meeting, seem to be the only action possible in the circumstances. We are determined that no mere war shall deflect us from our ultimate aim, but we realise that any attempt to carry on as we have done in the past would be, even if it were possible, largely a waste of effort. Unofficially those officers who are able to do so will continue to help the Society in the interplanetary ideal whenever they can. We hope to carry on with the technical meetings in London, but naturally it is impossible to foresee, how long they will be able to continue.

When the war is over, it should not be too difficult to get the Society going again. The name will be there, the money, the organisation and knowledge, and, we hope, the officers and members.

It now only remains to wish all our members all good fortune in whatever they may be doing now and in the future. We hope that before long we may be writing to them again in happier times, beneath less threatening skies. Remember that whatever this war may do to our lives, it is only an incident in the History of the World, and less than that in the greater story of mankind that lies before us.

Summary of Events Aftermath - 1937 to 1939

- By April 1937, the transfer of the Society complete
- After the first nine issues the London *Bulletin* reverts from its printed format to that of a single quarto sheet
- The *Bulletin* edited successfully by Ted Carnell, A Clarke then Maurice Hanson
- New, larger issue of the *Bulletin* to be edited by A Clarke and two other officers (Bill Temple and Maurice Hanson) would assist
- The new Hon Treasurer, Arthur Clarke rounds up the Liverpool members for their subscriptions
- Details re William Temple
- "The British Fan in His Natural Haunt"
- The second British Science Fiction Convention is held in London 10 April 1938
- Professor A Low voted as President of the SFA
- Arthur Clarke, William Temple and Maurice Hanson, share the flat at 88 Gray's Inn Road.
- 88 Gary's Inn Road becomes combined headquarters of the science fiction and interplanetary travel in London
- William Temple replaces Edward John Carnell as Publicity Director of the BIS
- Edward Carnell becomes editor of *New Worlds* and partner with Johnson in Science Fiction Service
- Six out of 30 turn up at the combined Liverpool BIS/SFA meeting called for 16 April 1937
- Appeal in *Novae Terrae* for Liverpool enthusiasts to rally around
- Professor Low offers a cup for best model airplane using jet propulsion
- Over 18 persons (including Gillings and Cleator) attend BIS/SF meetings in Liverpool and on 19 August 1937
- Also present was Cleator's fiancée Miss Birmingham
- Cleator informed that *Seeker of Tomorrow* had been published in *Astounding Stories* July 1937

- Gillings becomes editor of *Tales of Wonder* Summer 1937
- *Transatlantic Rocket Mails*, by L Johnson published in the *Meccano Magazine*, September 1937
- *Satellites of Death* by L Johnson, published in *Tales of Wonder* Summer 1938
- Science Fiction Service, Ted Carnell and Room 7 at No. 15 Houghton Street
- The Liverpool branch of the Science Fiction Association
- Meetings of the Liverpool branch formed of the SFA
- Meeting of the Liverpool branch of the SFA established at 15 Houghton Street
- *Science Fantasy Review*
- Names of pre-war BIS/SF Liverpool Hall of Fame
- *Fantasy* appears, edited by T Stanhope Sprigg and published by Messrs Newnes
- "The Tower Restaurant" marks the site of No. 15 Houghton Street
- The *Journal* of the BIS published in London one issue in which the years 1937, 1938 and 1938
- Technical projects to be undertaken by the London group
- Details re London version of the *Bulletin* edited by A Clarke.
- The August 1939 (last) issue of the *Bulletin* edited by Clarke
- Hilda Margaret Crossen and myself married 26 August 1939 - depart on honeymoon to Llandudno
- Hitler Invades Poland Britain declares war on Germany 3 September 1939.
- At an emergency meeting on 10 October 1939 the BIS is suspended for the duration

AFTER ARMAGEDDON

I am pleased to relate that to the best of my knowledge all those with whom I had been acquainted in the related fields of interplanetary travel and science fiction had survived the war, at least physically; whether we had survived mentally and emotionally was another question

The Army and the RAF appeared to have claimed us in more or less equal numbers; James Free was amongst the first away, with the Army, even before the war started, as he had been a sergeant in the territorials. Wally Gillings had experienced only a very brief Army career while Ted Carnell and Bill Temple ended up in the Mediterranean area, as I did myself. Despite a long search of Algeria and Tunisia for me by Ted Carnell, we did not have the pleasure of meeting.

In action with Central Mediterranean Force, Temple lost two written versions of his pre-war story, *The Three Pylons*, together with the manuscript for an 80,000-word novel, which was an expanded version of a previously published story. Also to the Army went Professor A Low to work on "experimental stuff".

To the RAF went Arthur Clarke to become a Flight Lieutenant, working on the Ground Controlled Approach Radar System, Norman

Weedall, to become a bomb armourer mainly in Burma, while Eric Frank Russell and myself became wireless mechanics. Russell divided his time mainly between Limavady (Northern Ireland) and the BAOR and complained that for three years he had always been separated from home by a boat.

He had probably less reason to complain, however, than I had had myself as the last three of my near six years of war were spent in North Africa, Italy, Corsica, the South of France and Italy again. On the way from Naples to Corsica by infantry landing craft, I probably passed Bill Temple, as he lay on his Anzio beachhead. I finally returned to Britain in the middle of December 1945, travelling by "Medloc" train from Naples to Dover. In fact, after three weeks' leave at home, I was due back at midnight on New Year's Eve at Hednesford RAF Station, Staffordshire in order to be demobilised. Such were the subtleties of the RAF, but needless to say, I did not arrive there until late the following afternoon. I wasn't the only one in my contingent to return late - much to the futile fury of the Officer-in-Charge.

Having heard that the BIS was out of hibernation, I wrote on 17 January 1946 to the enrolment secretary in order to get myself up to date on our relationships, and was very pleasantly surprised to receive a letter in return from my old friend, Arthur (Flight Lieutenant) Clarke, then stationed at the RAF station Atherstone, Warks., who had been shown my letter by Lionel Gilbert, while he was visiting him. Incidentally, I suppose a quite amusing sight would have been that of Corporal Johnson RAF, throwing a salute at Flt. Lieutenant Clarke RAF!

Arthur stated that as soon as he became demobbed he intended to go to London University to take a degree in Astronomy, and to this end, he was taking his Inter in July. In case he could not get any grant from the Ministry of Labour, he had written ten stories in the last two years, and had sold every one of them.

Not for the first time in my experience (nor the last!) Arthur stated that he was to retire from the writing field, even before any of his stories had been published, and was busy typing the 42,000 words of his "final" masterpiece. He went on to say, in a letter dated 6 March 1946, that both Carnell and Gillings had been involved in the forthcoming science fiction magazine, but "as he had stopped writing, he wasn't much interested."

In response to my queries, Arthur went on to say that with regard to the old BIS and CBAS (Combined British Astronautical Societies) that all previous members now started from scratch. Everyone interested in the Society would have to join in, like an ordinary member - and "even Low coughed up his guinea like a lamb". With regard to science fiction, there existed a very vocal minority in the new Society, who would not touch SF with a "barge-pole" and the new *Journal* would have to be as free of SF influences as the pages of "Nature".

This was, of course, the first time I had ever heard of the Combined British Astronautical Societies (having been away in the RAF overseas for three years) and it appears that in spite of wartime conditions, a new Society, The Astronautical Development Society had been founded in 1941 by K Gatland (whom I well knew as a member of the London branch of the BIS) and H Pantlin.

I had been aware of the existence of the Manchester Astronautical Association under Eric Burgess, and the late Cyril Cusack, this association having resulted from a schism from the Manchester Interplanetary Society; this Society after affiliating with a BIS was voluntarily disbanded shortly before the outbreak of the war. It was with the Manchester Astronautical Association that the Astronautical Development Society had amalgamated in early 1944 under the title of the Combined British Astronautical Societies with K Gatland as Secretary and Eric Burgess as Chairman.

Full details of the moves that resulted in the further amalgamation of the CBAS and the BIS are given elsewhere- notably in the issue of the BIS publication, *Spaceflight* dated July 1967.

But briefly - on the very day that I was due back at Hednesford RAF Station at midnight 31 December 1945 - the new National Society, The British Interplanetary Society Ltd received its certificate of registration as a company limited by guarantee (and not having a share capital). The Presidency, formerly held by Professor A Low was left in abeyance until "the honour which the Society wished to be associated with this post could be justified" in the characters of those to be considered for the position.

While Arthur Clarke, uniquely, was able to combine science fact and science fiction successfully, the idea of interplanetary travel (encouraged by the arrival of V1 and V2 from Germany) had become respectable and had generally to be separated from the harridan of science fiction that had given it birth and meaning. Not everyone had the ability to combine the best of both worlds as could Arthur Clarke.

Carnell, Gillings, Russell, Temple and myself largely went back to our first love, science fiction. Carnell beat me to it by producing the professionally printed *New Worlds* in April 1946, under Pendulum Publications, including in Vol 1 No 2, a short article that I had written called "Ahead of Reality". I, myself, beat Gillings to it by producing my magazine, *Outlands* in October, while Gillings edited *Fantasy*(not to be confused with Stanhope Sprigg's *Fantasy*) in December 1946, under the aegis of Temple Bar Publishing Company.

Temple went back to his "detested" stock exchange and Russell to his travels on behalf of Fredrick Braby and Company, and both went back to writing science fiction. Russell accomplished this with such success that he was eventually able to retire early from Braby's and concentrate on writing.

James Free and myself returned to our chores in the Liverpool Education Department, from which we both retired at the age of 60 years respectively. On account of distribution difficulties, my own publication, *Outlands* had failed after the first issue. Norman Weedall returned to organise his firm of window cleaners, and to become a well-known science fictioneer and frequenter of science fiction conventions.

Time marched on, and the 21st anniversary of the BIS was to be celebrated in Manchester by the north western branch of the Society. Unfortunately, something went wrong with the arrangements that were to have been made to invite to the gathering such of the original

Liverpool members as may still be available. As it happened, I was able to be present as a guest of my old friend, Norman Weedall, and there I was able to meet at least one former acquaintance in the shape of R Smith, who had been Organising Secretary of the London branch before the war. As Norman and myself were busily quaffing large whiskeys, we tended to ignore one of the speakers at the dinner, who (for whatever reason) had been distinctly uncomplimentary towards the early Liverpool members of the Society.

In later years, I was very pleased pleasantly surprised to receive a letter dated 16 July 1968 from Messrs Metro Goldwyn Mayer Pictures Ltd to state that Arthur Clarke had asked them to extend to me a cordial invitation to see the picture, "2001:A Space Odyssey" together with a friend, when the picture of which Arthur had written the script was shown at the Abbey Cinema, Liverpool. I was most pleased that Arthur had remembered me after so many years, and so it came to pass that Jimmy Free and myself were able to enjoy the film as Arthur's guests.

I had resumed acquaintance with Colin Askham after the war, having accidentally encountered him one day in Cases Street, Liverpool, near Headquarters of Messrs Littlewoods Mail Order Stores. John Moores had taken Askham into the firm as his personal secretary. Although the matter of the transfer of the Society to London had not been entirely forgotten, Colin was later good enough (as a Director of Everton Football Club for many years) to present Norman Weedall and myself with cup final tickets, whenever we had the urge to attend at Wembley - and on condition that the tickets were for our own personal use

And it was to Askham's bungalow in Formby (then in Lancashire, but now in Sefton District of the Merseyside County Council) that my wife and myself were invited one Sunday afternoon 20 July 1969, in order to witness (or at least to listen to the broadcast) when the lunar module, The Eagle, landed on the surface of the moon. Also present was Colin's wife, Eileen Hastie, who had been a pre-war member of the BIS and who had taken the Chair during at least one meeting of the Society during that period.

The television presentation of the lunar landing was not due to take place until between midnight and dawn on the Sunday night/Monday morning. Accordingly, on arriving home from our visit to the Askham's, I set up the television in the bedroom. When 12:30 am arrived with no sign of action and with the prospect of a day's work ahead of me (and especially a Monday's work!) I soon fell asleep.

I woke up during the night, and squinting at the alarm clock saw that the time was 3:55 am. I decided that Neil Armstrong must have emerged from the lunar module by that time, and switched on the set to see at what stage the proceedings had arrived. I was astonished to hear the announcer state that in five minutes Armstrong would emerge from the lunar module. So, in spite of myself, I was not to be denied. Almost as if by some miracle of thought transference, I had woken up just in time to witness an event that I had not expected would have taken place during my own lifetime.

Unhappily (and inevitably) the ranks of the pioneers of this science fiction and interplanetary travel, as I knew them are being thinned by the hand of time. Ted Carnell died suddenly on 23 March 1972, Eric Russell went suddenly at the age of 73 years on 28 February 1978; also, to depart this world was my old friend, Norman Weedall, who passed away on 19 August 1978 after a long illness, probably attributed to his war service in Burma. And a few days ago, I learned that Walter Gillings had died following a sudden heart attack on Tuesday 17 July 1979 at the age of 67 years.

Colin Askham, who was one of the oldest of the Founder Fellows is still with us at the age of over 80 years, together with his wife, Eileen. James Free, in his retirement, has been doing one day a week job as a courier for Littlewood's Pools (a job which I also had for a few months in 1976). It is possible that James Davies is still available, and the last I heard of Phil Cleator was that he had departed from thoughts of the future of mankind, and had turned to the past, writing books on archaeology.

Although the original British Interplanetary Society was officially wound up on 8 December 1945, it is perhaps an act of faith, that the new organisation retained the name of the old, despite the many attempts that have been made over the years (and that are still being made) to call the Society something else.

It is nice to see that the traditions of the Society, going back to the year 1933 are being honoured. Today, the Society is an institution of scientists, engineers, technologists, with a membership of 3,000 and is respected internationally. The Society publishes a highly technical *Journal* every month as well as a monthly edition of its more popular publication, *Spaceflight*. Its expansion is such that it has recently moved into new premises specially prepared to meet its ever-increasing administrative needs.

That I have lived to witness a lunar landing is something that I still find hard to believe. I scarcely expected this feat to be complete before the end of the century. It is a far cry back to Friday 13 October 1933 and HC Binns' office in Dale Street Liverpool.

Already the BIS is preparing to celebrate the 50th anniversary of the Society in 1983, but I wonder how many of us will be left to share in it.

The progress of the Society can perhaps be exemplified by the recollection that one (now well known) member of the Society, who in 1934 considered it too expensive to pay 2s 6d (12½p) for 100 sheets of members notepaper recently donated the princely sum of £10,000 towards the Society's Development Fund.

And where, I would like to know, are all those millions who (in spite of Isaac Newton) were convinced that a rocket would not work in the vacuum of space *"because it would have no air to push against ..?"*

Summary of Events After Armageddon (from 1945 to Apollo 11)

- The Army and the RAF had claimed the science fiction and interplanetary travel enthusiasts
- William Temple lost his MS in North Africa
- L Johnson demobilised on 1 January 1946
- Flight Lieutenant Arthur replies to the letter of enquiry re the BIS
- Clarke to take a degree in Astronomy
- Clarke wrote his "final" 42,000-word "masterpiece"
- All previous members of the BIS to start from scratch with the new Society
- Details concerning the combined British Astronautical Societies
- The British Interplanetary Society Limited was formed on 31 December 1945
- The Presidency of the new Society to be left open
- Science fiction and interplanetary travel face separation
- Only Arthur Clarke could combine the best of both worlds
- Carnell, Gillings, Russell, Temple and Johnson go back to science fiction
- James Free and Johnson return to the Liverpool Education Offices
- Norman Weedall returns to window cleaning
- The twenty-first anniversary of the BIS
- Johnson goes to the twenty-first anniversary dinner in Manchester as a guest of Norman Weedall
- Arthur Clarke, in July 1968, invites Johnson and friend to view "2001: A Space Odyssey" at the Abbey Cinema, Liverpool
- Acquaintance renewed with Colin Askham - and free Wembley tickets for Johnson and Weedall
- Colin and Eileen Askham invite Les Johnson and Hilda to a bungalow in Formby to "witness" the lunar landing, Sunday 20 July 1969

- Johnson awakes at 3.55 am in time to see the lunar landing - by the grace of Those Above
- Carnell, Russell, Weedall and Gillings all pass away
- P Cleator writes on archaeology
- The current position of the British Interplanetary Society
- Looking forward to the fiftieth anniversary of the BIS
- Long standing member of the BIS contributes £10,000 to the Development Fund
- There are those (anti-Newtonians) who were convinced that a rocket would not work in a vacuum? ***"There would be no air to push against"***

EPILOGUE BY GURBIR SINGH

The founding of the BIS and other rocketry societies was a product of their time. At the time, the world was going through profound social change triggered by fundamental discoveries in science, technological innovation, and social and political upheaval. In the early 1930s, scientists understood the structure and behaviour of subatomic particles for the first time. Cinema-goers began to hear people on the screen, not just see them move. The BBC was founded in 1922, allowing millions of people to share a common experience simultaneously. Amateur radio blossomed around the globe, enabling individuals separated by thousands of miles to communicate instantly. The advent of scheduled commercial flight, a phenomenon that had been the realm of magic, became an everyday reality.

1933 was not a good year for the arrival of the BIS. Three years earlier, Britain had suffered a dramatic and fatal failure of a new innovative aerial transport. On the night of 3rd October 1930, the airship R101 crashed during its maiden flight in the French countryside, killing 48 of the 54 people on board. It involved a far more significant loss of life than the tragedy that would strike the Hindenburg in 1937. The catastrophic disaster of the R101 was not recorded on film. The crash not only killed the airship's key designers and engineers who built it but sealed the fate of the airship program in Britain. Built at the same time, the R100 had completed a successful journey to and from Canada. Despite its unblemished record, it was sold for scrap. Britain terminated its airship program after an accumulated cost of about one million pounds.[1] For a government recovering from one disastrous adventure in new aviation technology (airships), keeping its distance from another (rockets) was at that moment in time a logical decision.

World War Two started six years after the BIS was founded. All activities stopped, and many of its members were deployed in military service, most overseas. Phil Cleator suffered a series of illnesses before, during, and after the war. His ill health informed the medical assessment that kept him from serving in the Forces. As a key port city of the British Empire, Liverpool was the most heavily bombed city in England outside London. Cleator suffered. The same German Luftwaffe raid severely damaged his house in March 1941, destroyed his family house nearby, and killed his parents.[2] In the several temporary dwellings he was forced to live in during the war, one house in rural Wirral had no running water or electricity.

Why did Cleator not re-join the BIS after the war? His persistent health problems, traumatic experience of the war, his abrupt communication skills (being perceived as a dictator), standing down from the BIS presidency, and the critical public reviews of his 1936 book *Rockets Through Space* contributed to him keeping away from the BIS once it reformed in December 1945. Most significantly, he decided to leave Britain in 1949. To recover from the physical and psychological agony of six years of war, Cleator and his wife moved to South Africa between 1949 and 1953. This was the period when the BIS played a vital role in establishing the International Astronautical Federation and elevated its national and international reputation and influence. The widening distance between Cleator and the BIS during this period probably ended any hope of him making any meaningful contribution to the society he had founded. Cleator wrote around 20 books in his lifetime, but only three were associated with space travel. Most were on history, ancient languages and archaeology. He wrote at least three more, including a play that remains unpublished. He passed away in 1994.

Leslie Johnson returned to Britain after the war in December 1945 as corporal Johnson. He was born at the beginning of the First World War and became a dad near the beginning of the second. He did not meet his first daughter until she was four years old. His formal demob at RAF Hednesford in Staffordshire on the 1st of January 1946 coincided with the launch of the reconstituted BIS Limited in London. He contacted the BIS in January 1946 and was told that everyone would have to re-join as an ordinary member. The former president, Prof. A. M. Low, had done just that, paying a subscription of one guinea. With the main BIS activities taking place in London and the demands of a young family, Johnson felt remote as he was distant. He kept an eye on BIS activities from Liverpool but never re-joined. His partnership in the mail-order service Science, Fiction Service with Ted Carnell, ended in 1945. He established Millcross Book Service, which he operated at least into the 1960s.[3]

Trevor Cusack, a founder member of the Manchester Interplanetary Society, was killed in an attack at sea in 1942. Stanley Davies returned to Manchester from Dunkirk but died soon after from injuries he sustained there.[4] Harry Turner, the editor of the MIS journal and a prolific space artist, was posted to India as a radar mechanic. His return was delayed until 1946.[5] As an instructor on radio and electronics for the RAF, Eric Burgess, was posted in rural England, so he was able to continue burning the interplanetary movement flame through the war.[6]

Arthur C. Clarke, who initially failed to get into the RAF because of the "absences of teeth and presence of spectacles"[7], eventually joined the RAF and worked on a Top-Secret project that allowed aircraft to land safely using radar in poor visibility. William F. Temple, who had shared the flat in London with Clarke, returned safely from Italy and established a successful writing career. Following a short career with the army, Walter Gillings also returned to writing after the war. Born in 1905, Eric Frank Russell was one of the oldest initial BIS members. According to Leslie Johnson, who had published *Seekers of Tomorrow* jointly with him in 1937, Russell had worked as a wireless mechanic for the RAF based in Northern Island and later in Germany with the British Army of the Rhine. Like Temple and Gillings, Russell turned to writing sci-fi as a career.

The war took Leslie Johnson, Ted Carnell and William F. Temple to the Mediterranean region. Still, according to Johnson, despite trying to meet during their time overseas, they did not make contact until their return after the hostilities concluded. Johnson was in Italy when the war ended and returned to Liverpool in December 1945. He remained with the Liverpool Education Offices until he retired aged 60 in 1974. He never returned to the heady days of writing and publishing fiction of the 1930s. His numerous and demanding roles with the BIS, presentations at the Science Fiction Association and writing articles (fiction and non-fiction) were all behind him. He started this manuscript in 1974 and completed it by the autumn of 1979. He sent copies to Arthur C. Clarke in Sri Lanka and historian Frank Winter at the National Air and Space Museum in Washington DC, intending to publish it on the 50th anniversary of the BIS in 1983, but sadly he passed away in July 1982. Of the many tributes that followed, Arthur C. Clarke wrote, saying, "he was very kind to me in the 1930s, patiently answering innumerable schoolboy questions. A few years ago, he wrote his memoir and sent me a typescript which is now of considerable historical value".[8] The BIS's 80th anniversary was commemorated in Liverpool in October 2013. His daughter Pam Reid presented a paper summarising her father's contribution.

Johnson never built or tested rockets. He did not conduct scientific research nor contribute to the invention of technology linked to spaceflight. From the outset, his deep curiosity focused on the intellectual possibilities of future societies was expressed through writing. In a piece he wrote in 1937[9] entitled "*Transatlantic Rocket Mail*", he concludes that for inanimate cargo ", transatlantic transport can be accomplished in 15 minutes". This vision of point-to-point rocket transport is now being actively investigated by commercial operators such as SpaceX, Blue Origin and Virgin Galactic. A Chinese company has announced winged rockets that will transport people, not just cargo, using "high-speed point-to-point transportation". The company Lingkong Tianxing based in Beijing, is working on a timetable for a prototype by 2024 and an operational service by 2030.[10]

The scale of death and destruction in 1942 motivated the young authors in the spaceflight community to turn inwards and reassess the political and social structures that determined the quality of life on Earth. In the pages of sci-fi magazines and science journals, they debated the meaning and relative values of democracy, socialism, capitalism, superstition and religion. They speculated on the characteristics of the future world order. In a piece, Harry Turner wrote, "War can only be avoided by the creation of a world state", John Burke visualised a "world state where disputes could be handled without war and injustice", and Arthur C. Clarke envisaged that by the early 1960s a "scientific world state" would be in place.[11] There was also opposition. The idea of humans perfecting transport technology to leave the Earth and venture to new worlds in the Solar System filled the eminent writer C. S. Lewis with foreboding and revulsion. Lewis considered the ambitions of the interplanetary community to be amoral, dangerous and ready to "open a new chapter of misery for the universe. It is the idea that humanity, having now sufficiently corrupted the planet where it arose, must at all costs contrive to seed itself over a larger area".[12] Clarke attempted to encourage Lewis to engage with the BIS multiple times, but he kept his distance. Perhaps because of that distance, Lewis had a broader, more realistic view of the interplanetary movement and where it could lead. In 1954, Arthur C. Clarke with Val Cleaver met with C.S. Lewis and J.R.R. Tolkien in an Oxford pub to debate their opposing views. Clarke recalls that "neither side converted the other, and we refused to abandon our diabolical schemes of interplanetary conquest".[13]

Post-war events suggest that Lewis and not Clarke had been nearer the truth. Despite the sincere desire of the interplanetary community, the accelerated development in rocketry was the product of a nationalistic political ideology in pursuit of military conquest. In the absence of the rocket's capacity to offer a military advantage, it would not have attracted the substantial financial support that ultimately made the space age possible on the timescale it happened. The rhetoric of space superiority and the fear of falling behind kept the investment coming and sustained the Cold War.[14] Many of the early BIS members wrote about the exploration of the Solar System and humans travelling to the Moon but did not expect to experience it in their lifetime. Wernher von Braun is one of astronautics controversial figures. The only one to have received a medal from Adolf Hitler and President Eisenhower (in 1944, the Knight's Cross for developing the V2 and in 1959, the Distinguished Federal Civilian Service Award from President Eisenhower).[15] He portrayed himself as a reluctant scientist forced to develop rockets for military use when all he wanted was to explore space.

Writing in July 1945, in part to ingratiate himself with the Americans (to whom he had chosen to surrender), indicated that his military work on rockets was not yet done. He could develop a spacecraft to monitor troop movement, deliver nuclear weapons from a spacecraft in Earth orbit, and build a giant mirror (hundreds of km in diameter) in Earth orbit. Power from space could be redirected to any place on the Earth or even change the local weather.[16] The BIS's 1965 position paper on the future of space in the United Kingdom more than hinted at the military potential of a national space programme. This Faustian compact has accompanied human progress throughout history in all its endeavours. It was no surprise that the first steps towards space travel would be motivated by nationalistic and military goals.

Most of the early BIS members were writers. Post-war members included Carl Sagan, Eric Burgess, Olaf Stapledon, George Bernard Shaw, Robert Heinlein and Patrick Moore. Clarke, Robert Heinlein, Willy Ley and Isaac Asimov were considered "Astrofuturists".[17] They were all scientifically literate, and scientific principles guided their writings. Many of the visions they imagined then underpin our 21st-century society today. For example, the focus of Clarke's celebrated 1945 paper, *Extraterrestrial Relays*, is now deployed in communication satellites that provide communication services anytime, anywhere between anyone on Earth. In 1971, Clarke was invited by the US State Department for the signing of the 80-nation INTELSAT

(International Telecommunication Satellite Organisation) agreement. He concluded his speech by saying, "you have just signed far more than just another intergovernmental agreement. You have just signed a first draft of the Articles of Federation of the United States of Earth".[18] By chance, in 1975, when he was living in Sri Lanka, Clarke was one of the first to have a private satellite link at home.[19]

In his 1973 novel, *Rendezvous with Rama*, Clarke imagined a human society living in colonies amongst the solar system's planets. Rather than a World State, he imagined a United-Nations-like body but for all the colonies of the Solar System. He called it United Planets and based its headquarters on the Moon. Johnson, Cleator, Clarke and all their contemporary writers were innately optimistic and shared principles of Humanism. In October 1939, Cleator described war as the "supreme and ultimate imbecility of the human species".[20] Rather than undermining their pre-war naive desire for a utopian future for humanity, their first-hand experiences of war vindicated their belief in a peaceful future for a united human race empowered by spaceflight.

The interplanetary community wrote about a future in which their hopes and wishes of the 1930s would be an everyday reality. A utopian vision where space technology could deliver education, fresh food, medical needs, social interactions and intellectual fulfilment. Much of the imagined technology has arrived, but these benefits are not yet equally distributed to all the people on the planet. That technology allows more people to live longer, healthier lives for the first time in human history. Still, poverty, discrimination and gross inequality persist within and between nations.[21]

An unassuming young man from Liverpool, born in the year the First World War started, Leslie Johnson not only lived through a remarkable period in history but also made a personal contribution to it. As a teenager, he established the Universal Science Circle. He does not write about the thought processes that led him to create it. He did not explicitly express his aspirations for a peaceful world united by a single international language powered by modern technology. But all his contributions convey that vision. It is a vision far from being realised, but incremental progress towards it continues to be made.

. . .

Gurbir Singh

April 2022

---

[1] Singh, G, India's Forgotten Rocket Pioneer. More details on this program are in the section entitled Airship. P65.

[2] Phil Cleator describes his wartime experience in Mencken, H. L. (Henry Louis), 1880-1956, Letters from Baltimore: the Mencken-Cleator correspondence. The above publication was edited by Cleator and published in 1981.

[3] Prolapse, Fanzine for Temporal Regression, August 2008. p38. The name "Millcross" came from "Mill Lane", his address and "Cross", his wife's maiden name.

[4] Trevor Cusack was reported to have been "Killed at sea through enemy action". Astronautics - Volume 12, Number 5 3 October 1942.

https://arc.aiaa.org/doi/abs/10.2514/8.10341?journalCode=jastn

Stanley Davies died in August 1941 from injuries he endured at Dunkirk (via an interview with his daughter Ann)

https://astrotalkuk.org/episode-54-23-july-2012-manchester-interplanetary-society-and-stanley-davis/

[5] No longer on a war footing, he found time to visit a cinema in Bangalore where he noted the *"Quit India' movement has a strong following here, yet I find the locals decidedly friendly on these excursions"*.

Harry Turner returned from India in 1946. His son Philip maintains a website with much of his father's material, including paintings and notes from his time in India.

http://www.htspweb.co.uk/fandf/romart/het/history/india03.htm

[6] Eric Burgess outlines his wartime activities in the preface of his 1991 book, Far Encounter, The Neptune System. Burgess dedicated this book to Phil Cleator for convincing him that "spaceflight could be changed from a dream to a reality during our lifetimes."

[7] This comes from a letter dated 17/09/1940 from Clare to Simon You'd, editor of a British sci-fi magazine Fantast.

Clarke Archives. Smithsonian Air and Space Museum. Washington DC.

[8] Clarke, A.C, Spaceflight Vol 25 July/Aug 1983. The manuscript in question is this one.

[9] September 1937, The Meccano magazine Vol XXII No. 9

[10] Despite the pandemic, 2021 was one of the most productive years for space missions coming out of China. In addition to the space run CNSA, China is home to numerous private-sector space companies. Linking Tianxing is one such private company. . https://www.cnbeta.com/articles/science/1227257.htm

[11] Fantast was produced by C S Youd; 14 issues were printed between 1939 and 1942.

Fantast July 1942 - Arthur Ego Clarke ("ego" was a nickname Arthur acquired when living at 88 Gays Inn Rd.) A Short History of Fantocracy 1948-1960

Fantast March 1940 Harry Turner, Creed of an Atheist.

Fantast Sep 1939, John F Burke

[12] This is an extract from C. S. Lewis's 1943 novel Perelandra. A few months after it was published, Clarke wrote to Lewis

[13] Clarke, AC., 1993. By Space Possessed. P18

[14] Neufeld, Michael. 2006 Space superiority: Wernher von Braun's campaign for a nuclear-armed space station, 1946–1956. Space History Division (MRC 311), National Air and Space Museum,

Smithsonian Institution. https://repository.si.edu/bitstream/handle/10088/29811/Space%20Superiority.pdf

[15] I should say that von Braun is the only one I am aware of. If you know of others - do let me know.

[16] Braun, von Wernher, June 1945, Declassified by the UK Minister of Supply. Published in JBIS March 1951. Survey of development of Liquid rockets in Germany and their prospects

[17] de Witt Douglas Kilgore, 2003 Astrofuturism: Science, Race, and Visions of Utopia in Space.

[18] Clarke, A.C. "Profiles of the Future", London, 1962. P213

[19] Singh, G, The Indian Space Programme. P155. India cooperated in an educational programme called SITE. Using NASA's ATS-6 communication satellite to bring educational TV programmes to rural Indian villages. By chance, the satellite's footprint included Siri Lanka.

[20] Letter dated 20/10/1939 from Cleator to Menken.

Letters from Baltimore: The Mencken-Cleator correspondence by Mencken, H. L. (Henry Louis). They were published in 1982.

[21] Stephen Baxter, who had worked with Arthur C. Clarke, wrote a tribute to him in 2017. In that piece, he explores Clarke's visions of utopia.

Communications from Utopia: Sir Arthur C. Clarke, Science Fiction, and the United States of Earth, Journal of the British Interplanetary Society, Vol. 70, pp.426-429, 2017

PICTURES

Leslie Johnson 1940. Credit. Private Collection of Leslie Johnson

46 Mill Lane, Liverpool. Where Leslie Johnson lived until he married and where the Universal Science Circle met. Credit. Private Collection of Leslie Johnson

Arthur Charles Clarke in his study 1935. Private Collection of Leslie Johnson

Walter Gillings. January 1935. Editor of Britains first Scifi magazine Tales of Wonder. Credit. Private Collection of Leslie Johnson

Norman Weedall. Credit. Private Collection of Leslie Johnson

www.ingramcontent.com/pod-product-compliance
Lightning Source LLC
Chambersburg PA
CBHW021145080526
44588CB00008B/228